D0443749

Dr. Beach's Survival Guide

Selected Other Books by Stephen P. Leatherman

AAA Beach Vacation Travel Journal
America's Best Beaches

Dr. Beach's
SURVIVAL GUIDE

What You Need to Know About Sharks,

Rip Currents, and More Before Going in the Water

Stephen P. Leatherman
Director, Laboratory for Coastal Research
Florida International University

Yale University Press   New Haven & London

Copyright © 2003 by Stephen P. Leatherman.
All rights reserved.
This book may not be reproduced, in whole or in part, including illustrations, in any form (beyond that copying permitted by Sections 107 and 108 of the U.S. Copyright Law and except by reviewers for the public press), without written permission from the publishers.

"Dr. Beach" is a registered trademark of Stephen P. Leatherman. All rights reserved.

Designed by Sonia Shannon and set in Adobe Garamond with Torino and Futura display type by Integrated Publishing Solutions. Printed in the United States of America by R. R. Donnelley & Sons, Harrisonburg, Virginia

The paper in this book meets the guidelines for permanence and durability of the Committee on Production Guidelines for Book Longevity of the Council on Library Resources.

Library of Congress Cataloging-in-Publication Data
Leatherman, Stephen P.
Dr. Beach's survival guide : what you need to know about sharks, rip currents, and more before going in the water / Stephen P. Leatherman ("Dr. Beach").
p. cm.
Includes bibliographical references.
ISBN 0-300-09818-9
1. Bathing beaches—Health aspects. 2. Bathing beaches—Safety measures. 3. Shark attacks. 4. Tides. I. Title: Doctor Beach's survival guide. II. Title.
RA606 .L38 2003
613′ .12—dc21   2002013849

A catalogue record for this book is available from the British Library.

10 9 8 7 6 5 4 3 2 1

To my parents, Evelyn and John Leatherman,
for all of their encouragement for my research endeavors, and
especially for taking me to the beach when I was a child

Contents

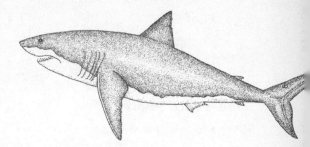

Preface

In the summer of 1995, Jim Kennedy and his family made their annual pilgrimage from Boston to Ocean City, Maryland, arriving just before dusk. Ten-year old Thomas was eager to get in the water, so he and his father went to the beach for an end-of-day swim. They didn't know that the lifeguards had closed the beach that day because of the high surf kicked up by Hurricane Felix; the guards were off duty by 6 P.M.

Father and son walked along the sand until they found an area of lower-than-average waves. The boy was only up to his waist in the water when he felt the strong tug of a rip current. His father grabbed his hand and, anchoring his feet in the shifting sand, tried to pull Thomas to shore. But the current was too strong. The boy's hand slipped out of his, and Jim Kennedy watched in horror as his son was swept away to the deep ocean waters; there was absolutely nothing that he could do to help. The U.S. Coast Guard recovered Thomas's body floating near Ocean City Inlet.[1]

The boy and his father had made a classic mistake. Not knowing how to spot the tell-tale signs of a deadly rip current, they had entered the water where it appeared to be the safest—exactly where this offshore-moving current was literally knocking down the waves.

I was interviewed by the *Baltimore Sun* and the local press to explain how this terrible mistake could have been made. The father later called me and said he wanted to do something to warn others. He wanted to install signs and diagrams along the Ocean City boardwalk that described the dangerous rips, but the city managers were against this sort of awareness campaign. They maintained that rips are rare occurrences, and that the signs would only serve to scare people and business away from their beach.

This sorrowful episode spurred me to write this book.

With more than half the U.S. population living within 50 miles of the coast, and the majority of summer vacationers spending time at beaches, the importance of beach safety cannot be overemphasized. Beaches are seductive places, but they can be dangerous. The purpose of this book is to help make your beach visits safer by explaining the risks and hazards at that magical margin of sea and sand.

Acknowledgments

I have always been fascinated by sharks—superb creatures that have been around for more than 300 million years. My professional interest in sharks developed from my interaction with Dr. Eugenie Clark (the "Shark Lady") when we were both professors at the University of Maryland, College Park. Genie shared her love and insight for sharks with me as we were planning some studies together in the Red Sea. She founded the Cape Haze Marine Laboratory, which is now the Mote Marine Laboratory (MML) in Sarasota, Florida.

This book profited from my discussions with Dr. Robert Hueter, director of the MML Center for Shark Research, the world's largest scientific program focusing on the biology and behavior of sharks. Much of the data on shark statistics are from George Burgess, director of the International Shark Attack File. Ms. Marie Levine of the Shark Research Institute assisted with some photographs.

The biggest danger at ocean beaches is drowning in rip currents—hazards that have touched me personally. All beaches should be overseen by lifeguards certified for ocean rescues. I take my hat off to the men and women of the United States Lifesaving Association (USLA).

I would also like to thank Susan Lichtman, Harald Johnson, and Heidi Downey for their editorial reviews, which greatly improved this book. My wife, Debbie Leatherman, and Carolyn Walker toiled through many early drafts, and their patience and assistance is much appreciated. Professional reviews were provided by Tobey Curtis and Dr. Robert Hueter (for sharks) and Jim Lushine (for rip currents and lightning strikes). I appreciate the many insightful conversations that I had with these scientists about the subject areas that they are so knowledgeable about and love.

It has been a pleasure to work again with Jean Thomson Black of Yale University Press after the great success of our first book together (*Barrier Islands,* Academic Press) several decades ago.

Introduction

The sun dipped toward the horizon at the small beach town of Avon, North Carolina. Most beachgoers along the Outer Banks had long since packed up their towels and gear on this fateful day of September 3, 2001. But one couple remained for a last dip in the enticing sea. The Russian nationals from the Washington, D.C., area were swimming together near a sandbar just a dozen yards offshore when a shark suddenly attacked them. The young man, Sergi, 28, sustained massive injuries to his lower extremities, including the mangling and loss of most of one leg. The foot of his companion, Natalia, 23, was severed, and her buttocks and hips were severely wounded. The water filled with blood as their screams rang out. Onlookers pulled the gravely injured pair to shore, but it was too late for Sergi, who went into cardiac arrest and died on the beach. Natalia survived, but she needed extensive plastic surgery and a long recovery period. The beach had claimed two more victims during a summer that had the media in a shark frenzy.

Beaches are the most popular recreational areas in the United States; more than 200 million Americans trek to the shore each year. Beaches help fulfill our desire to return to nature. When families reminisce, they often recall their beach vacations as being among the most memorable. Yet ocean beaches can also be hazardous, and few people understand or know how to recognize potential trouble.

Although gruesome shark attacks do occur, sharks are statistically not a great danger. On average, only one person a year is killed by sharks in the United States. Far more people are killed by bee stings, bathtub falls, and lightning strikes than by shark attacks. The real threat to swimmers and bathers at beaches are rip currents, which cause an estimated 70 to 100 drownings annually. Yet most people do not even think about rips, much less fear their power or know how to recognize them.

When I was a boy, I couldn't wait to get to the beach. And my favorite way of entering the ocean was to run as fast as I could into the water until it was too deep and I fell forward. This showed my enthusiasm for the ocean but also my ignorance of its power.

I am now a scientist—a professional coastal geomorphologist—and I study beaches around the world. In fact, as Dr. Beach I annually rate and rank the best beaches according to 50 variables that I've developed in my years of visiting and analyzing the coast (see appendix E). In the following pages I will take you on a tour of the beaches I know so well and point out the creatures and the safety risks that go along with them. My goal is to alert you to the dangers so that your wonderful beach vacation does not turn into a tragedy.

one**Sharks**

It was the summer of 1916, and optimistic Americans were celebrating their inventiveness and prosperity. On July 1, 25-year-old Charles Vansant went for a long swim off the New Jersey shore at Beach Haven. As he was heading back to shore, someone spotted a large fin slicing through the water. The shark attacked Vansant, who died of his injuries. Five days later and 45 miles north, Charles Burder was swimming in the ocean near Spring Lake with a group of friends when a shark suddenly struck and severed his legs. He bled to death. The people of Matawan, New Jersey, a village 11 miles inland, seemed removed from the fears of their beachgoing neighbors. But on July 12 their sense of safety was shattered. Twelve-year-old Lester Stilwell was playing in the local watering hole, which was actually a small tidal creek. He suddenly screamed and disappeared beneath the water. Several men jumped in to save him. Stanley Fisher, one of the would-be rescuers, found the boy's mangled and lifeless body and was pulling it to shore when he, too, was attacked by something. Fisher died of his injuries. Further down-stream, a group of boys were swimming in the same creek and unaware of any danger. Someone started yelling for the boys to get out of the water. Fourteen-year-old Joseph Dunn was the last boy out of the water; he was climbing the ladder to safety when his leg was seized by a shark. Dunn lost his leg but escaped with his life.

On July 14, a nine-foot great white shark was netted at South Amboy, New Jersey. The shark's stomach contained fifteen pounds of human flesh and bone, including the shinbone of a boy and a human rib.[1] The final devastating tally? Four dead, one seriously injured. During this Twelve Days of Terror occurred the most gruesome series of shark attacks in American history. Nineteen-sixteen became the Year of the Shark, and, as a result, hundreds of sharks were caught and slaughtered along the mid-Atlantic coast. This compelling story of a marauding shark was the subject of the best-selling book *Close to Shore* by Michael Capuzzo, who writes, "People who are attacked by sharks are exceptionally, almost absurdly, unlucky."

Although the number of deaths from such common incidents as bee stings or falling off of ladders is far greater than from shark attacks, many people have an almost hysterical response to shark attacks—the thought of being torn apart by a shark is ghastly indeed. Just mentioning the word shark conjures a mixture of fear and fascination in the human psyche—we never seem to hear enough about these creatures, which are often regarded as monstrous killing machines. The public has had a bad case of shark phobia since 1974, when the book

Jaws, by Peter Benchley, was published. But the myth of shark attacks often surpasses the reality.

There are more than 400 species of sharks. They range in size from the one-ounce pygmy shark to the 28,000-pound whale shark, the largest fish in the ocean. Like almost all sharks, the whale shark is harmless to humans—it is a filter feeder that consumes plankton and small fishes.

Few people ever have a shark encounter, even though scores of sharks are often swimming just offshore. While sharks generally are not waiting to attack, there is always the potential for trouble, especially for someone who spends a considerable amount of time at beaches and in the water. My friends and colleagues and I have had our share of shark encounters of the very close kind. Here is a sampling, along with some other insights about shark behavior:

In 1998, I consulted for the rebuilding and revitalization of Lucaya Beach on the Grand Bahama Islands. This resort had deteriorated physically; the beach was badly eroded with part of the seawall falling into the water, and tourism was at low ebb, to say the least. My job was to evaluate the beach's condition and to locate sources of sand for use in restoration. As part of the preliminary investigations I snorkeled far offshore, studying the ocean bottom and diving down to examine pockets of sand that might be used to replenish the beach (and that were far from the living coral reefs). I never even thought about sharks as I was conducting my explorations up to a thousand feet offshore. (It's a tough job, but someone's got to do it!)

Fast forward to August 2001 and imagine my reaction when I heard about a man who, while swimming in the same area of Lucaya Beach that I had surveyed, and in only five feet of water, had his leg bitten off by a large shark. The Long Island, New York, vacationer lost so much blood from the massive injury that he was lucky to be alive. Considering the severity of the attack, it had to be a big shark, but the swimmer never saw it coming in the stirred and murky water created by Tropical Storm Barry.

I have seen sharks most often in tidal inlets, where the water is deep and the currents swift. These are the places you expect to see sharks, as there is great movement of nutrients and fish through these watery connections between the open ocean and the nursery grounds of bays, lagoons, and sounds. I once taught a summer course at the Duke University Marine Laboratory, and my students and I often took the boat *Privateer* to the outer barrier islands, especially Cape Lookout. While we cruised through the inlets we occasionally spotted lone sharks in the six-foot range, their fins cutting the water. It was always exciting to come close to these fearsome creatures while in the protective sheath of a boat.

Dr. Eugenie Clark, a.k.a. the Shark Lady, is one of the most heralded shark experts of all time. She once told me about a fisherman who was bitten by a large hammerhead shark off the coast of the Red Sea. The poor man showed Genie how this monster shark had severely bitten his leg. What was confounding was that the attack had occurred in only four feet of water—very shallow water indeed for a predator almost ten feet long. Actually, the man had caught the shark and pulled it into his boat, thinking that he had killed it. But the shark was still alive, and it bit him. All of this had, in fact, happened in four feet of water, but the incident did not occur as it was portrayed by the media.

At one point in my career I directed a coastal research center at the University of Massachusetts in Amherst. While most of my time was spent conducting research at Cape Cod National Seashore, I also spent time on the islands and at many other New England shore areas. Once, while sitting on the dock at Duxbury in Massachusetts Bay on a sunny fall day and enjoying my lunch, I started thinking about going for a swim as I threw scraps of my sandwich to the appreciative gulls. My feet were dangling over the edge, not far from the water's surface. Suddenly, huge jaws came straight up from the water beneath, opened wide, and engulfed one of the large herring gulls that was sitting on the calm water. The long, bony teeth closed over the entire body of the struggling sea gull except for one of its desperately flapping wings. The wing continued to flutter in the toothy mouth as this predator slowly sank into the water, never to be seen again. My first thought was that this was a shark attack, but I later learned that it was the work of a large goosefish (called monkfish at the fish market), which is partial to gulls but perfectly harmless to people.

I grew up in the Carolinas, and one of my favorite beaches has always been Cape Hatteras. This is an area that I visited many times as I paid my way through college conducting beach erosion surveys. This spectacular ribbon of sand constitutes the Outer Banks of North Carolina—a long chain of barrier islands jutting into the Atlantic Ocean. When the surf was not too big I liked to swim far offshore into water tens of feet deep. Of course, I was younger and stronger then, so swimming hundreds of yards offshore wasn't a problem.

During one of my swims on the south side of Cape Hatteras, my upper leg was hit so hard that it jolted me in the water. I was sure that it was a shark. I had heard stories that big sharks can take off your leg with one painless bite. As I slowly moved my arm down to examine my leg, my hand came in contact with the flat nose of a huge loggerhead sea turtle. When she popped up to the surface for some air I got an up-close view of this giant sea creature, which must have weighed 400 to 500 pounds. She was most likely waiting for nightfall before coming ashore to lay her eggs; she bumped into me by mistake. I was relieved, to say the least, but this experience made me think hard about the wisdom of swimming far offshore all alone in deep ocean water.

I regularly travel along the U.S. coasts for research and to conduct my beach surveys, performing my duties as Dr. Beach. I particularly like to visit Florida's beaches during spring and fall to avoid the crowds, obtain good accommodations for a reasonable price, and still enjoy the warm water.

Florida has the clearest water on average of any U.S. continental beach because the base rock is limestone, and there are no rivers or streams carrying silts and clays to muddy the ma-

rine waters. I was once walking along the beach conducting my
survey near Fort Pierce Inlet when I noticed a darkish area in
the otherwise clear blue water. At first I thought a large patch
of seaweed was washing ashore. As I neared the blotchy area,
which measured many yards wide, I realized that I was approaching a huge school of bait fish. These finger-sized fish
were moving right up on the shore, and some were jumping out
of the water onto the dry beach. At first I tried to rescue the fish
by throwing them back into the water. Then it occurred to me
that fish don't normally jump onto beaches . . . something must
be scaring them into this behavior. Larger fish or small sharks
had to be feasting on the schooling fish. Anyone who happened
to be in the water during this feeding frenzy would almost certainly be bitten. I alerted the lifeguard whose stand was a few
hundred yards down the beach. He immediately hoisted the
danger flag for no swimming.

Spearfishing commonly attracts sharks. The Florida Keys is a
favorite area for this activity because of its clear tropical waters.
For spearfishermen, the harmless nurse shark can be a real nuisance. You have to literally fight off these sharks as they try to
snatch a free meal with their small, stringy teeth. While nurse
sharks are not aggressive, you can bring on an attack yourself.
A few years ago a teenager who was spearfishing grabbed the
tail of a three-foot nurse shark as it swam by. The shark swung
around and bit the boy's chest so tightly that it stayed attached,
through its ability to form suction, until removed by doctors at
the nearby hospital in Marathon Key. The doctors were forced
to split the shark's spine to unlock its jaws.[2]

The really big and potentially life-threatening sharks are often circling just out of sight. If you are scuba diving and one of these predators wants you, there is not much you can do about it. Sharks hit like a freight train—the attack is over before you even have time to react. The scariest time for divers is when surfacing. This is when sharks like to take a shot, coming up quickly from the depths. More than 90 percent of shark attacks occur in near-surface water.

Before I moved to South Florida, I made several trips to Miami Beach in the late 1970s to evaluate the erosion problem and propose solutions. At that time, waves lapped at the hotel seawalls at high tide; there was little to no beach, and tourists were scarce as well. Then came the most massive beach restoration project in the history of the world up to that time. More than 13 million cubic yards of sand were pumped from the offshore sea bottom through a pipeline to create a new beach over 9 miles long and 300 feet wide. The tourists returned in droves, and Miami Beach today is again considered a world-class beach.

Just before the beach restoration project started, a group of coastal scientists and engineers (including me) visited Miami Beach for a fact-finding mission. After a full day of meetings, some of us wanted to really experience the beach first-hand. It was December and the days were short, but the water was still warm, so we went swimming in the dark. There were no lifeguards to warn us, but we should have been smarter, since sharks often come out to feed at night.

Fortunately, no one was attacked on this outing, but we later learned that a large school of blacktip sharks up to six feet long had been spotted a few days earlier in the same waters.

Next time I will recommend that we hit the hotel pool and save the beach swimming for daytime.

The irregular swimming actions of animals tend to attract sharks. People who swim in shark-infested waters with a dog greatly enhance their chance of being attacked. In 1987 a man and his poodle went swimming from his boat near Panama City, Florida. Within minutes a large bull shark struck, tearing at the man's legs; he died shortly thereafter in the bloodied water. The International Shark Attack File contains many accounts of sharks drawn to human victims by the erratic motions of a paddling dog.

Some shark researchers now believe that the phase of the moon is fundamental to shark behavior—full moons are credited with having a powerful effect on all living things. Interestingly, the gruesome attacks and "shark rampage" of 1916 occurred during an eclipse of a full moon. There is evidence that sharks attack more frequently during very high tides, corresponding to times of strong gravitational pull of the full moon. According to George Burgess, the noted shark expert, sharks could be reacting directly to the moon or, most likely, to its effect on other ocean species—the reproduction of corals and many types of fish is affected by moon phase. Of course, very high tides during full moons flood the beaches, resulting in deeper water closer to shore and perhaps drawing in seals and sharks.

Several of my friends are avid fishermen who like to fly to the Bahamas to catch large groupers and sailfish. Some years ago Dan Tuckfield was making one such flight back to Fort Lauderdale with a fishing party when the small plane developed engine problems. The people on board jettisoned all they could to lighten the load, but the twin-engine Cessna still could not stay aloft. The pilot had to let the plane drop into the Atlantic Ocean, where it sank to the bottom within an hour. All five people survived the crash, but some passengers could not swim well and, worse, became panicked, splashing and flailing when they saw tiger sharks.

The sharks, averaging nine feet in length, swam circles around the people, but they did not attack. An older man swallowed too much water and drowned. As his body sank, the sharks immediately consumed the corpse. The man's wife soon suffered a similar fate.

Tuckfield, being a strong swimmer, was able to swim 17 miles to Cat Cay, south of Bimini—a 40-hour ordeal in the open water. When the U.S. Coast Guard was dispatched to rescue the remaining two survivors, seven sharks were swimming around them. Obviously the tiger sharks could have attacked at will, but they instead chose to be scavengers rather than marauders.

The Unholy Trinity

Three species of sharks are responsible for nearly all of the serious (life-threatening) attacks in the world: great white, tiger, and bull sharks (figure 1). Of the 980 shark attacks on humans and boats between the years 1580 and 2000 as recorded in the International Shark Attack File, more than half were attributed to these three: great white (348), tiger (116), and bull (82).

Figure 1. The Unholy Trinity: Great white sharks (top), temperate to cool waters; average length 15 feet, average weight 1,500 pounds. Tiger sharks (center), subtropical waters; average length 11 to 14 feet, average weight 850 to 1,400 pounds. Bull sharks (bottom), U.S. East Coast, Gulf of Mexico, some freshwater rivers; average length 7 to 8 feet, average weight 285 pounds.

Many of the others were nonserious bites by small sharks. (The numerous shark attacks during the summer of 2001 near New Smyrna Beach, Florida, were attributed to blacktip sharks.)

While the great white shark is generally found in cool waters, the tiger shark is a tropical species. It is known as the denizen of the deep off the Hawaiian Islands, and it is usually seen only when hooked by fishermen who are purposely shark fishing. Tiger sharks do not venture into shallow water unless they are enticed by blood in the water or are on the trail of easy prey, such as dead, floating fish. Great whites and tiger sharks will often attack the torso of a human victim, which is bite size for such large predators.

The bull shark is the least commonly known of the "problem" sharks. Although they are not nearly as large as great whites or tigers, they can do tremendous damage to a leg or an arm (an appendage that they can get their jaws around). Bull sharks have been implicated in a number of deaths in recent years in a series of vicious attacks along the U.S. mid-Atlantic and Gulf coasts. The bull shark is the most dangerous shark in U.S. waters, except for the great white in the Red Triangle of central California.

Let's take a look at these sharks, which pose the greatest threat to beachgoers.

Great White Sharks

Just as the jaguar is the most fearsome member of the feline family because of its head size, the huge and fierce great white is the deadliest of the sharks. It can reach a length greater than twenty feet, weigh several thousand pounds, and have an upper jaw perimeter of more than four feet—that jaw reveals rows of razor-sharp serrated teeth more than two inches long.

Great white sharks are the sea creatures that inspired Peter Benchley's best-selling book *Jaws*. Benchley wrote about, and the big screen portrayed, a shark of prehistoric dimensions— perhaps 25 to 30 feet long. While super sharks of this size did exist tens of millions of years ago, there has never been a modern-day great white shark that reliably measured longer than 21 feet (about 6.4 meters). Benchley's story and the *Jaws* movies set off a wave of shark-attack hysteria across the country. Our fear of and fascination with sharks has never declined.

The great white shark, known by Australians as "white death," is one of nature's most efficient killing machines, with sharp teeth and a mouth large enough to engulf a person's entire body. In fact, great white sharks have on rare occasion completely consumed people; this is the stuff from which legends are made. The sight of their soulless black eyes and of their prominent triangular dorsal fin knifing through the surface water is chilling. We can admire these great hunter-killers, but we must do so from a safe distance: great white sharks merit the same caution we would accord tigers, lions, and bears, for they are dangerous to all other animals, including humans. But like grizzly and polar bears, great whites are found in particular areas—you have to be at the wrong place at the wrong time to find trouble.

When fully grown to 20-foot goliaths, great white sharks are the masters of the sea—there is virtually nothing for them to fear. Other sharks scatter when a great white approaches. According to Edward O. Wilson in *The Diversity of Life,* the great white shark is "with the saltwater crocodile and Sundarbans tiger the last expert predator of man still living free . . . by all odds the most frightening animal on Earth, swift, relentless, mysterious, and unpredictable." The signature of a great white attack is its suddenness and ferocity; sometimes the prey is hoisted right out of the water as the shark attacks from beneath, racing to the surface at full speed.

While large great white sharks have been caught in deep water off Montauk Point, New York, and elsewhere, California is the only state to record any attacks in recent decades. Since 1950 there have been 79 attacks in California waters, 7 of them fatal. The victims generally are diving for abalone or surfing in an area populated by seals and sea lions, the sharks' favored prey. It is risky business diving for abalone in shark-infested waters, which are primarily the underwater jungles of kelp off California's coast. Even professional divers have been killed. Encounters with great whites are rare—the Surfrider Foundation reports that there is only one shark attack for every million surfers.

Most attacks by great whites have been recorded in the so-called Red Triangle—a 100-mile stretch of coastal waters bounded by Bodega Bay to the north, the Farallon Islands to the west, and Año Nuevo (just north of Santa Cruz) on the south. Skin divers and surfers alike use wet suits in these cold waters, and these black silhouettes can look like a seal meal to a hungry shark.

Most of the attacks occur in August and September, when the warmer water draws great numbers of swimmers and surfers to beaches along the California coast. There have been a number of shark attacks at Stinson Beach in Marin County, just north of San Francisco. This large pocket beach, bounded by rocky headlands, is a favorite for boogie boarding in the big Pacific Ocean swell. Nearly every year a great white shark chomps down on a surfboard, probably mistaking it for the underside of a surface-swimming seal. Generally the surfer gets some lacerating wounds to the thigh or other nasty bites but is able to paddle back to shore. A local newspaper reported that a 10-foot-long great white "narrowly missed breakfast" in September 2000 when a 34-year-old surfer escaped unharmed. The teeth marks, 13 inches apart at their widest on the surfboard, made a nice imprint of the bite. Paul Euwer later posed for photographers with his souvenir surfboard. His comment to the press was that a close call with a shark isn't a big deal. "It's a calculated risk you take," he said.[3]

Humans are fortunately not a favored prey of sharks. Great whites do not search us out for food; rather, they mistake humans for their normal food of seals and sea lions. They are merely "taste testing" wet-suited swimmers and surfboard riders who resemble seals from below. When a number of kayaks or surfboards are in the water, it is usually the short ones that are attacked. This suggests that sharks are looking for the silhouette of the most vulnerable seal or sea lion, or are targeting the general size of their preferred prey.[4]

Great whites need the high caloric content of fatty animals, like seals (fat constitutes almost half the total body weight of juvenile elephant seals). That is why great whites rarely come back for a second bite after tasting human flesh—we are just too bony and scrawny, and eating us is wasting storage space in their stomachs. Fat has twice the energy value of muscle, critical for great white sharks. They burn a lot of energy swimming in the chilly waters off central California and Australia (two of their favorite haunts) because they have a higher metabolism than most sharks and are partially warm-blooded. Of course, one good bite from such a large predator can obviously be fatal (figure 2).

Great white sharks are such key players in the coastal ecosystem and food-web balance that they are a protected species in most U.S., Australian, and southern African waters. However, shark conservation is problematic given the public perception of great whites being *Jaws*-type monsters.

Peter Benchley has now joined the chorus of ocean conservationists trying to protect great white sharks from human predators, who kill them for sport or a type of revenge. Benchley, who comes from Nantucket, points out that far less was known about great whites when *Jaws* was written in 1974. "Back then, it was generally accepted that great whites ate people by choice. Now we know that almost every attack on a human is an accident: the shark mistakes the human for its normal prey."[5] Nearly three-quarters of all bite victims survive because the great whites recognize their mistake and don't return.

Tiger Sharks

Tiger sharks, beautiful animals with black stripes on their back, are definitely the most dangerous sharks in tropical waters. They can reach lengths of 18 feet, and 10- to 15-footers are reg-

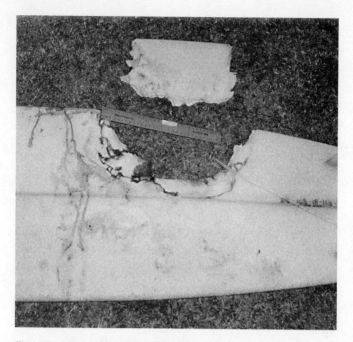

Figure 2. A great white shark clamped down on a surfboard. Courtesy Global Shark Attack File, Shark Research Institute.

ularly seen off the Hawaiian reefs. These opportunistic predators eat a variety of marine mammals and seabirds, preferring
sea turtles. Tigers are one of the few sharks that actually con-
sume people, and they will eat almost anything when hungry,
including other sharks (even smaller tiger sharks), rays, various
bony fish, and garbage from passing ships. The stomachs of
killed tiger sharks have been found to contain such unusual
items as dogs, beer bottles, boots, and unopened cans of beans.

Sharks are well known in the deep waters that closely surround the Hawaiian Islands. In some spots only a few hundred
feet offshore the seaward front of the shallow outer reefs drops
off into thousands of feet of water. In the Carolinas, where I
grew up, you would have to go hundreds of miles offshore to
find water even approaching these depths. During mid-August
2000, I was attending a professional coastal conference in
Maui, when a 15-foot tiger shark was spotted at one of my top-
rated beaches—Kapalua Bay Beach, on the north end of Maui,
the first National Winner in 1991.[6] Japanese fishermen had been
trawling the deep waters just offshore and were throwing the
junk fish back into the water. The current and prevailing wind
carried the dead fish into the protected, shallow, and crystal-
clear waters of Kapalua Bay Beach. The lifeguards immediately
spotted the shark as it swam closer to shore, gobbling up the
dead fish. No one was attacked, but can you imagine being that
close to one of these frightening creatures?

Attacks by tiger sharks are often associated with feeding.
These sharks are often attracted to stream mouths, especially
after heavy rains, when upland fish and other food sources are
swept out to sea. Such murky waters are prime feeding areas,
and you don't want to be in the water when a tiger shark is in
feeding mode. Another dangerous place to be diving or surfing
is in an area populated by sea turtles. While it is fun to watch
and swim with sea turtles, many of the recorded attacks in
Hawaii have occurred in such areas. There is a sea turtle hatch-
ery offshore of Hanalei Bay on the island of Kauai, and large
tiger sharks are often spotted in these waters. I watched skin

divers entering the water from the long pier at Hanalei Bay; they were hoping to swim with a turtle and perhaps see a shark "at a safe distance." Surfers have been attacked by tiger sharks because of mistaken identity—the bottom of a boogie board looks like the underside of a turtle from a shark's perspective. There have been some nasty bites and sometimes loss of hands or feet, but most surfers are able to get to shore.

A ten-foot tiger shark attacked Jesse Spencer while he was surfing the Kona coast on the Big Island of Hawaii in October 1999.[7] The shark came halfway out of the water and pushed the teenager off his board; it then proceeded to lock its jaws onto his arm. Miraculously, Jesse managed to free his bleeding arm as the shark bit down on his surfboard. Jesse's mistake was trying to catch that last wave at sundown; tiger sharks are generally night hunters.

Tiger shark attacks on humans are increasing in Hawaiian waters. One obvious reason is the increasing number of people swimming and participating in water sports. Another reason may be that the turtle populations have rebounded in recent years because of various protective policies. Green turtles come close to shore for feeding, and the tiger sharks follow their prey. Unlike great whites, which take their prey by surprise, and bull sharks, which ambush their victims, tigers are slower, and people often see them coming. If you are bitten by a tiger shark, get out of the water as soon as possible, as they are likely to come back for the rest of the meal.

Tiger sharks are carnivorous scavengers, looking for an easy meal. Stimuli that can excite their feeding behavior are speared and bleeding fish or injured and struggling humans, particularly if there is blood in the water. Tiger sharks will normally remain in fairly deep water except to follow such prey.

While the tiger shark that swam into Kapalua Bay was easily identified by lifeguards in the shallow, clear waters, encounters in deeper water by surfers are less certain. An identifying feature of such an attack is the serrated teeth of the tiger shark, which can be found lodged in a surfboard or its victim. Shark teeth are made to break off—when one does, a new tooth rotates to the front.

Tiger sharks have been implicated in many of the attacks in the warm Pacific waters following a downed plane or a sunken ship. During World War II a great number of servicemen, both Japanese and American, died in shark attacks in the bloody waters. Many men staved off the sharks by shooting and kicking them until they could be rescued, but the soldiers often sustained some injury. Movies often depict this kind of situation, with the sharks swirling around their intended human prey.

Bull Sharks

Bull sharks generally grow up to ten feet long and can weigh 500 pounds, but they make up for their relative lack of size in aggressiveness—they are considered the most pugnacious of all sharks. Scuba divers and spearfishermen say that the bull shark is, pound for pound, the most dangerous animal in warm waters. According to Robert Hueter, director of the Center for

Shark Research at the Mote Marine Laboratory in Sarasota, Florida, bull sharks have the highest level of testosterone known in any animal, including lions and elephants.[8] The bull shark, an ambush type of predator, will attack prey as large as it is. It is the only shark that poses a grave threat to people swimming in shallow water, especially close to tidal inlets and channels.

Because of their varied habitat and relative abundance, bull sharks are more likely to have contact with people than are the better-known man-eaters—great white and tiger sharks. Bull sharks are found from New York to Florida and in the Gulf of Mexico, as well as off South Africa and elsewhere. Unlike most sharks, they range from salty ocean water to the brackish water of the Chesapeake Bay, and have even been caught hundreds of miles upstream in the fresh water of the Mississippi River. Although smaller than the other two members of the Unholy Trinity, the bull shark has massive jaws studded with large teeth fully capable of inflicting fatal wounds to humans.

Bull sharks were largely unknown to the public until the shocking mauling in 2001 of a boy at a beach in the Florida panhandle. And while it is not known which shark species caused two separate and deadly attacks along the Virginia and North Carolina beaches in 2001, a prime suspect is a bull shark.

In early July, eight-year-old Jessie Arbogast was enjoying the shallow surf at the Fort Pickens area of Florida's Gulf Islands National Seashore, which is close to a major channel, Pensacola Pass. Jessie was reportedly playing with his friends only 15 feet from shore when he saw a large fin approaching. The shark first bit his arm and then took out a chunk of his thigh.

The little boy screamed to his uncle on the shore to "get him off me."[9]

By the time the uncle reached Jessie, the 6.5-foot, 200-pound bull shark had nearly bitten off and was trying to swallow the boy's right arm. The muscular uncle, who trains for triathlons, grabbed the shark by its sandpaper tail and managed to pull it up on the beach. Jessie's arm had become detached, and the boy had been nearly drained of blood. The shark was thrashing violently. A National Park Service ranger shot the shark four times in the head and then was able to pry open its mouth to recover Jessie's severed arm. The boy was rushed to the hospital by helicopter, where he lay unconscious as blood was quickly pumped back into his body. Jessie Arbogast survived this ordeal (with both arms). At the time of this writing, he is conscious but cannot talk or walk because of extensive brain damage.

Ten-year-old David Peltier was mauled by a shark while surfing off the Virginia coast during the 2001 Labor Day weekend; he could not be saved despite his father's efforts to free him from the shark's jaws with his bare hands. David was attacked at 6:00 P.M. while surfing on a sandbar in four feet of water about 50 feet offshore.[10] The 8-foot-long shark made a 17-inch gash on the boy's left leg, with bite marks extending from groin to ankle. The father fought back with his hands, punching, pushing, and kicking, but the boy bled to death, as the main artery in his left thigh was severed. The species of shark in the attack is not certain, but the savagery, the large size of bite, and the circumstances of the attack point to a bull shark. This is the case with most shark attacks; positive ID in these times of stress is usually not possible except when the attacker is a hammerhead shark—its unusual-shaped head can never be mistaken.

The second fatal shark attack (on the Russian couple described earlier) occurred the same weekend at Cape Hatteras along the Outer Banks of North Carolina.[11] The severity of the wounds and the nature of the attack clearly point to a predatory shark, most likely another bull shark.

Such back-to-back attacks in close proximity are extraordinary. In fact, the Virginia shark attack was the first reported death in about 30 years in those coastal waters. Although the odds of such an attack happening are slim, this spate of attacks in 2001 left many beachgoers afraid to go into the water. The attack on Jessie Arbogast had captured the imagination of the public and ignited a media frenzy about shark attacks that has not been witnessed since the *Jaws* era.

In August 2000, a retiree jumped off his dock into Tampa Bay for a swim. He landed virtually on the back of a shark, estimated to be a nine-footer weighing 400 pounds. The shark, which had been chasing and feeding on mullet, proceeded to rip Thadeus Kubinski apart while his wife screamed hysterically in the back yard.[12] The bay there is relatively deep due to a channel dredged for boats, and its waters are not as clear as those of the Gulf. In addition, the bay is subject to currents carrying nutrients and small fish. The bay water was stirring that day, likely the result of the shark chasing after or feeding on the fish. The splash of the victim probably triggered the attack. The crescent-shaped shark bite extended from Kubinski's right armpit to the hip—a total of 15 inches. This pattern, along with the triangular impressions left by the half-inch-long teeth, are characteristic of the bull shark.[13] People who swim in Tampa Bay have come close to these predators quite often, but they typically get away unharmed. Thadeus Kubinski was not so lucky, as he died shortly after this ghastly attack.

Shark Beach

Shark spotting is a daily occurrence along Florida's central Atlantic coast, and New Smyrna Beach, just south of Daytona Beach, is the shark attack capital of the world. The good news is that the bites are generally not serious, and no one has been killed in recent memory. These are small sharks feeding on schooling fish; blacktip and spinner sharks in the three- to five-foot range are often implicated in the attacks.

With over 70 million tourists a year, 1,350 miles of coastline—the most in the United States except for frigid Alaska—and 80 percent of its residents living and working near the water, Florida is the leader in shark attacks in the world (see figure 3). Of the 338 attacks that have occurred in U.S. waters since 1990, 220 were in Florida.[14]

South Florida is relatively free of shark attacks because of the narrow continental shelf; the Gulf Stream comes close to shore here. Because of the proximity of the deeply flowing Gulf Stream, there is little habitat along the famed Gold Coast from Palm Beach to Miami for the nearshore coastal sharks that lurk around the beaches farther north. Also, the southern waters are clearer than the rather murky coastal waters of New Smyrna Beach and the adjoining Ponce de Leon Inlet, so that

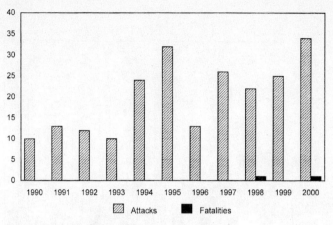

Figure 3a. Shark Attacks in Florida, 1990–2000

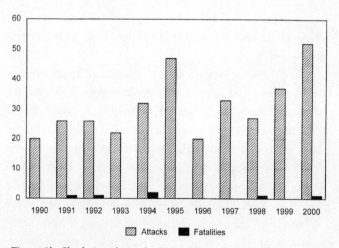

Figure 3b. Shark Attacks in the United States, 1990–2000

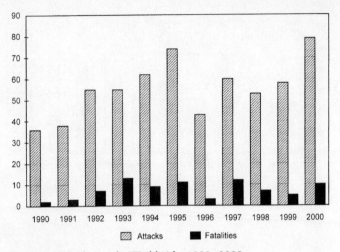

Figure 3c. Shark Attacks Worldwide, 1990–2000

sharks rarely mistake people's feet or hands for their preferred prey—fish.

New Smyrna Beach is one of the closest beaches to Orlando, the number one vacation destination in the world, the home to Disney World and other attractions. There are 47 miles of coastline in the vicinity (Volusia County), and thousands of people, mostly kids, love to surf the three-foot waves and swim in the warm waters. There were 22 shark attacks during the summer of 2001 off New Smyrna Beach (25 percent of the world total for the year), most of them near Ponce de Leon Inlet, one of Florida's best surfing spots.[15] None of the bites were life threatening. People wading in the murky water were bitten on the thigh, while surfers were often nipped on the legs or feet.

Although these small sharks are more a nuisance than a severe threat for swimmers and surfers, the attacks kept writers and TV commentators busy.

In late August 2001, officials closed New Smyrna Beach to swimmers and surfers for the first time in at least 30 years. Sharks had been biting at people's limbs almost daily for several weeks. The combination of murky water caused by heavy rains and thus considerable runoff into the ocean, a large population of bait fish, and a popular surfing contest had driven up the numbers of bites by small sharks mistaking humans for prey. When the water is clearer there is really no problem, even here at so-called shark-bite central.

The attacks at New Smyrna Beach are all about the food chain, and humans are not part of it—just in harm's way. The Ponce de Leon Inlet area experiences regular cold water upwellings from deep ocean water. These colder waters that reach the surface are filled with near-microscopic-size plant life (plankton). The plankton is consumed by millions of tiny fish, which are eaten by larger fish and so on up the food chain to the top predators—sharks. Surfers are drawn to these waters because of the good surf on the large deltaic-shaped bars caused by the strong ebbing tidal currents of Ponce Inlet. So it is obvious why some surfers, with their feet dangling off the boards into murky water full of bait fish and sharks, get into trouble. In fact, some young surfers consider it a badge of courage to go into the water with the sharks so visible. It is misleading to classify these incidents as attacks because the sharks are mostly just grabbing and letting go, but they are still officially logged as bites in the International Shark Attack File. We need a classification system

that distinguishes simple bites from real shark attacks, which
are caused by much larger species of sharks and can cause life-
threatening injuries.

The spate of shark attacks at New Smyrna Beach both re-
pulsed and lured beachgoers. Vendors cashed in on the town's
reputation and the media hype. Oils labeled "shark repellent"
and souvenir hats and T-shirts—some reading BITE ME and
sporting pictures of sharks—were popular during the shark-
mania. It was almost a carnival atmosphere, though the surfers
were unhappy that going into the water was illegal—a real
bummer for end-of-summer wave riding.

In the end, some visitors decided to experience just Disney
World and canceled their reservations at New Smyrna Beach.
Hotel and motel occupancy plummeted. The loss of business
prompted Florida Governor Jeb Bush to weigh in, saying that
the "amount of news coverage is disproportional to the problem
that we face." Sharks are not the problem in Volusia County;
rather, the problem is the public perception and the media
hype over the so-called shark attacks (actually just minor bites).

Profiling and Avoiding Shark Attacks

The increasing number of shark attacks seems to point to a
growing threat. There is great public interest in these attacks, as
people want to know what is causing them and what they can
do to avoid them. It is important to remember that the shark
attack rate per number of people in the water actually has not
increased over time. The simple fact is that there are more

people in the water, which increases the chances of a shark encounter, as demonstrated by the Florida, California, and Hawaii statistics (see figure 4). Coastal populations are swelling,
and there has been phenomenal growth in aquatic sports, especially surfing, kayaking, sailboarding, and boogie boarding.

When I was growing up in the Carolinas during the '60s,
people canoed and kayaked on rivers, but sailboarding and
boogie boarding were unknown. Board surfing is something
that teenagers in Southern California did. I never even considered going to Cape Hatteras for board surfing; now it is the
number one spot for this activity on the U.S. East Coast.
Sharks weren't really an issue, either; the attacks we heard about
involved great whites and scuba divers hunting abalone along
the central California coast, as well as people surfing along the
Australian coast before the shark nets were installed at popular
beaches.

There are three kinds of unprovoked shark attacks—"hit
and run," "ambush," and "bump and bite."[16] There is a separate category for provoked attacks involving sharks that have
been hooked by fisherman or otherwise captured. Attacks related to spearfishing and shark feeding are classified as unprovoked. Historically, there have been 48 deaths attributable to
unprovoked shark attacks in U.S. coastal waters, which is only
13 percent of a total of over 700 encounters.[17]

The most common type of shark attack is the hit and run,
which happens in shallow water, typically three to five feet
deep, at popular beaches. Small sharks, ranging in size from
two to six feet, bite a swimmer, bather, or surfer on the leg, arm,
or thigh, and then let go. The turbulence and water turbidity

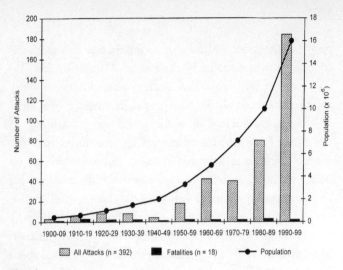

Florida

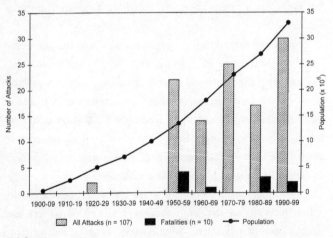

California

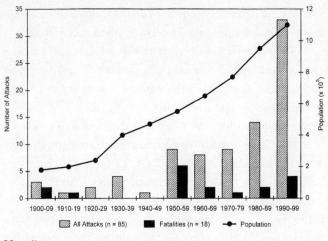

Hawaii

Figure 4. Shark Attacks vs. Population Growth in the Twentieth Century. *Source:* International Shark Attack File.

caused by the breaking waves, combined with the strong currents in the surf zone, necessitate quick strikes. This type of shark attack is more appropriately described as a shark bite; they usually occur when the water is the murkiest (following rains) and bait fish are plentiful. These conditions, combined with the fact that the sharks take only one bite and then dart away, indicate that such attacks are cases of mistaken identity. These hit and run attacks by blacktip and occasionally spinner sharks are most common along the U.S. Atlantic coast.

More severe injuries and occasionally deaths result from sneak or ambush attacks (great white and bull) and bump and bite or circling attacks (tiger). These incidents involve swimmers and divers and generally occur in much deeper water.

Bull sharks are probably responsible for most of the serious attacks on swimmers along the U.S. East and Gulf coasts. Most hammerhead sharks, even those in the five- to six-foot range, are not dangerous, although they look like they would be. Attacks by hammerheads are rare but unmistakable, so people report them with proportionally more certainty than those of other shark species.

Most shark attacks occur in warm waters. Florida leads the nation, averaging 17 attacks (mostly just bites) per year during the 1990s.[18] Next are Hawaii and California. Like Florida, California is a mecca for beachgoers, but it also has numerous surfers on short boards that, to a shark, can look like seals. Following California in the number of shark attacks is South Carolina, which has a long coastline, as well as somewhat murky waters (see table). Worldwide, Australia and South Africa join Florida in registering the most shark attacks, but those in the Southern Hemisphere are generally more serious because of the species involved.

About 90 percent of shark attacks are recorded in water depths of less than five feet. While the simple bites of small sharks occur in shallow beach water, the serious attacks by great whites and tigers occur in much deeper water—the victim is usually surfing or surfacing after a dive when the attack occurs.

The vast majority of sharks are not observed during the daytime; twilight and nighttime are their most active times for feeding. Obviously, this is best time for swimmers and divers to avoid the ocean. Feeding behavior varies by species, with bull

Shark Attacks in the United States, 1990–2000

State	Number of attacks	Number of fatal attacks
Florida	224	2
Hawaii	35	3
California	31	1
South Carolina	12	0
Texas	9	0
Oregon	7	0
Virginia	2	0
North Carolina	10	0
Georgia	3	0
Massachusetts	1	0
New York	1	0
Washington	0	0

Source: George Burgess, International Shark Attack File

sharks being the most problematic because of their sneak attacks. Because the ambush attacks of great whites are based on
mistaken identity, the best strategy of escape is to swim grace-
fully to shore; splashing and flailing around indicates injury
and an easy kill. Tiger sharks are more prone to attack divers
and swimmers in the presence of bleeding or dead fish; this can
be thought of as a feeding frenzy situation. Tiger sharks will
often circle their intended victim before initiating the attack;
the best strategy in this case is to get out of the water as quickly
as possible. Beyond these simple observations, there is no sci-
entifically documented pattern of attack because there are too
few attacks to detect any common features.

How to reduce your already slim chance of shark attack

• Never swim, surf, or dive alone; stay in groups, as sharks are more likely to attack a solitary individual.

• Never swim or dive with an open wound; sharks have an incredible sense of smell, and blood incites their feeding instinct.

• Never swim or dive during twilight hours or at night; most large predatory sharks feed at these times.

• Don't swim with pets; their erratic motions may attract sharks.

• Avoid dirty or murky water; many shark attacks occur because of mistaken identity—humans are rarely the intended prey.

• Do not wander too far from shore; this isolates an individual and additionally places one far from assistance.

• If a shark is spotted, swim as smoothly as possible out of the water; thrashing movements and wild splashing indicate a wounded animal and attract sharks.

• Avoid wearing shiny jewelry; the reflected light may look like shining fish scales to a shark at depth.

• Avoid waters with known effluents or sewage (as a general safety rule), which can attract bait fish and thus sharks.

• Avoid waters being used by sport or commercial fishermen, including surfcasters, especially if there are signs of bait fish or if there is feeding activity. Sea gulls diving into the water are an indication of such activity.

• Never grab an injured shark; many "provoked" attacks have occurred when a shark is hooked and brought onto the beach or into a boat.

• Don't harass sharks; even the normally placid nurse sharks can attack if grabbed.

• Sightings of porpoises do not indicate the absence of sharks; both animals often eat the same food and usually just ignore each other (contrary to mythology).

• Exercise caution when swimming, diving, or surfing near steep drop-offs; these are favorite hangouts for sharks.

• Don't swim near piers or jetties because of fishing activity (as well as the danger of breaking waves and rip currents).

• If attacked by a shark, do whatever is necessary to get away; some people have successfully thwarted attacks by hitting a predatory shark in the nose, gouging at its eyes or putting their hands in its gills.

• Swim or surf at beaches patrolled by lifeguards, and follow their advice.

Source: Adapted from the International Shark Attack File

Sharkmania and Misconceptions

Every summer sharks bite people; this makes news because of our fearful fascination with these predators. In Florida, the typical victim is bitten (that is, nibbled) on the leg or foot while bathing or paddling out in the murky waters off Ponce de Leon Inlet near New Smyrna Beach. Shark attacks such as the one resulting in the dismemberment of Jessie Arbogast or the deaths in Virginia and North Carolina are real news stories, but even these events are blown out of proportion. The summer of 2001 wasn't a shark frenzy; rather, it was a media frenzy with a season-long siege of media stories (during an otherwise slow news season).

NBC opened its nightly news coverage proclaiming the "Summer of the Shark," following the lead of a *Time* magazine story.[19] After the man in the Bahamas was mauled by a shark in early August 2001, the media went wild. The hospital where he was treated had to fend off interview requests from an enormous list of media programs, including: *Dateline, 20/20, Larry King Live,* and all the morning shows. A hospital spokesman compiled a list of media requests that was an astounding eight pages long—single-spaced! Only the end of summer and the tragic September 11 attacks on the World Trade Center and the Pentagon quelled this frenzy.

There is the perception by some people, certainly in response to all the media hype, that sharks are somehow out of control. Some people speculate that a change in environmental conditions, such as water temperatures, has been spurring more attacks. During the last century, sea surface temperatures increased by about one degree Fahrenheit in response to global warming. While this warming is already having an effect on certain animal life along the California coast (namely barnacles and other sea creatures that are attached to rocks), there is no evidence to suggest an influence on shark behavior. During El Niño years, water temperatures can soar by five degrees Fahrenheit or more; these years are the best for swimming without a wetsuit along the Southern California coast. Again, there was no surge in shark attacks for the major El Niños of 1982–83 or 1997–98.

Some sharks migrate seasonally in response to water temperature and food supplies. These migrations occur along the coast; schools of sharks numbering 20, 30, or more, are routinely spotted by helicopters or airplanes. Several years ago more than 100 blacktip and spinner sharks were schooling just off Miami Beach, but no one was attacked in the clear waters. A few days later, all the sharks were gone. Little is known about shark migration and the schooling phenomenon.

One of the most persistent theories of why attacks are increasing is that the shark's food supply has changed, perhaps dwindling because of overfishing. Therefore, sharks, as the top predators, are forced to scavenge close to shore, putting them into more frequent contact with swimmers and bathers. Overfishing is indeed a worldwide problem, but there is no evidence that it is altering sharks' feeding or attack behaviors. If changing food supplies were leading to more hungry sharks, then aggressively feeding sharks would do far more damage to people during attacks than has been recorded to date.

Shark fear has resulted in some people espousing the principle that the only good shark is a dead shark, and lunatics sometimes launch a personal vendetta by catching and slaughtering as many sharks as possible. The sharks are not rabid; it is people who are ignorant of the facts and are fearful.

One of the greatest misconceptions about sharks is that they generally attack people in order to eat them. Analysis of more than 1,000 recent shark attacks worldwide shows that well over half of these attacks had no relation to feeding. More than 75 percent of the victims were struck only once—sharks rarely pressed home their attack after the initial strike.[20] These statistics do not even include the bites and nibbles from small "sand" sharks.

The sea itself is far more dangerous than sharks, as an early 1980s story from Australia points out. During a single weekend, there was one shark attack; the person was not seriously injured. During the same weekend, 14 people drowned in the big surf and powerful rip currents for which Australian beaches are known. Inexplicably, the shark encounter garnered the greatest attention by the media.[21] I suppose the difference is that shark attacks are, by definition, spine-chilling events, while drownings are as mundane as boating accidents and traffic deaths.

Shark Feeding

A new and growing sport is feeding sharks when diving. Several years ago, while conducting some research in Nassau, Bahamas, I was invited by Stuart Cove to join a shark-feeding expedition. Cove takes divers down about 20 feet to the sandy bottom in the clear Bahamian waters on the largely unpopulated western end of New Providence Island. The divers kneel in a circle, and the sharks approach the guide, who lures them with cut fish for an exciting encounter.

I went along for the ride but had planned just to snorkel around and watch this activity from afar. When we arrived at the designated shark-feeding area, the clear and calm waters of the area were stirred up by the thrashing of sharks, their fins cutting the water. Most of the participants were alarmed by this spectacle. Robert Kennedy, Jr., of the famous Massachusetts family, put on his tanks and mask and jumped in, uttering something about how we only live once. Some followed him, but others got "wet feet" and decided to remain on board. The sharks followed the divers down to their standard feeding area, and there were no problems. Now such activities are frowned upon by environmentalists, including some people in this ocean adventure group.

In the wake of the shark-attack publicity in 2001, especially the horrific story of the swimmer who lost his leg in the Grand Bahamas, there were calls for a ban on shark feeding. Some feel that such a practice may encourage sharks to associate humans with food. While there are no confirmed reports of shark-feeding leading to shark attacks, Florida has banned all shark feeding in coastal waters. Dive operators (and some respected shark biologists, such as Robert Hueter) protest the bans, stating that such encounters, if conducted properly, are educational, not just thrill-seeking adventures. They further maintain that the ban is a result of the recent shark hysteria and that educating people about sharks will help avert the eradication of sharks.[22]

Some marine conservationists counter that shark feeding is not only unsafe but may be altering fish behavior and ecology. The bigger question, they point out, is: Should we be feeding wild animals? Except in zoos and aquaria, we do not feed bears, alligators, or dolphins. The ban on shark-feeding dives in Florida waters does not apply to the Bahamas and elsewhere, where these activities continue to be quite popular.

two**Rip Currents**

During a long, hot summer, nothing is more exhilarating than taking a plunge into the sea at your favorite beach. Never mind that the waves look a little larger than normal or that no lifeguards are on duty—the urge is almost irresistible. A big beach party celebrating Kwanzaa at American Beach, Florida, in 1994 ended in disaster when five people were pulled offshore to their death. The culprit was a rip current—a concentrated flow of water that can jet you offshore (figure 5). Unfortunately, these beachgoers did not know how to spot the killer currents, nor were they at a guarded beach. The group was completely unprepared to deal with these seaward rushes of water.

Most people are only vaguely familiar with beach hazards and the threats they pose to their safety. While coastal scientists have long studied waves and currents, the media are much less aware of the circumstances that can lead to drownings. Newspaper accounts of beach tragedies often lay the blame on nebulous causes, such as undertow, freak waves, or collapsing sand-

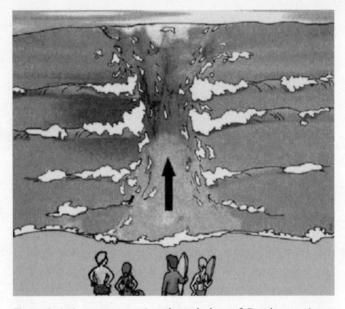

Figure 5. A rip current—a river through the surf. People sometimes look for areas where the waves are smaller and then go right into the rip current to their death. *Source:* Modified from U.S. Lifesaving Association manual.

bars. As Andy Short has pointed out, these catchy and alarming terms may have little relevance to the actual circumstances that contributed to the accident.

The Surf Beach

One of the things that makes ocean beaches so appealing and exciting is the surf—waves are the heartbeat of the ocean. We are drawn to this rhythmic pounding of the waves. I can spend hours gazing at the ever-changing shore. The fresh, salty air invigorates the body as the sheer beauty of the scene and the interplay between the waves and beach capture the imagination and refresh the psyche. Also, nothing is more fun than jumping into the water and playing in the surf, feeling the power of the waves crashing ashore. Unfortunately, the unpredictable nature of the surf and its power results in many injuries and deaths each year, even affecting experienced swimmers and surfers.

The 6,000 or so waves that strike a beach every day are generated by the wind. Generally speaking, as the wind speed increases, so does the surf. Waves that break on beaches can be generated locally or be spawned thousands of miles away by storms at sea. Hurricanes cause the largest waves, termed swell, along the Atlantic coast, while migratory low-pressure cells (storms) at high latitudes generate the great Pacific Ocean swells. Because distant storms are responsible for them, huge swell waves can hit beaches when the weather is perfect, sunny and cloud-free. The north shore of Oahu, Hawaii, is directly exposed to these giant ocean swells, which can reach 30 to 40 feet in January, the season of international surfing contests.

Wave height is the primary determinant of rip current strength, but wavelength is also significant. Wavelength refers to the width of the wave, measured from trough to trough. The height and width determine the volume of water in a wave. Some waves that peak when breaking may appear powerful, but there is no real force behind them without a large mass of water. I have sometimes been fooled while boogie boarding on the Carolina coast by these waves. By contrast, the big swells that dominate the Pacific coast tend to have long wavelengths, making them powerful waves that break with considerable force. While it is nearly impossible to measure wavelength when in the water, you can easily count in seconds the time between waves as they break. The longer the time between breakers (termed the wave period), the longer their wavelength and, consequently, the greater the force for a wave of a particular height. Long-period swell waves of 20 to 25 seconds are the best surfing waves along the Southern California coast, but these turbulent waters are best avoided for swimming; I suggest heading to the nearest heated swimming pool.

High waves can be very dangerous. What is not often understood by the public is that the energy they produce is proportional to the height of the wave squared. Therefore, a three-foot wave is nine—not three—times more powerful than a one-foot wave. When onshore breaking waves reach five feet, the surf is generally too dangerous for swimming. Experienced surfers look for the big waves, but good surfing beaches are often not safe for swimming.

Breaking waves can be classified into three primary types: plunging, spilling, and surging (figures 6 and 7). Plunging waves are by far the most exciting and dangerous, forceful and fast. These breaking waves are formed when swell suddenly encounters a shallow bottom, such as a reef, large sandbar, or steeply sloping beach. The wave is forced to peak and break suddenly, with all of its force concentrated in a limited area. Plunging waves often generate rip currents and shorebreaks on steep beaches and are responsible for many more injuries than spill-

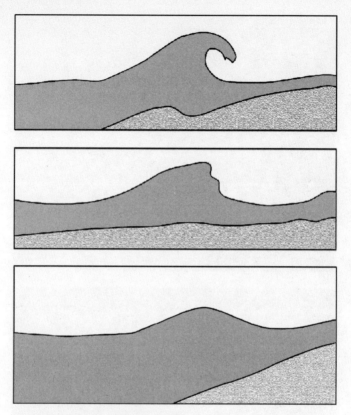

Figure 6. Types of breaking waves. Plunging breakers, top, are the most spectacular to watch. Spilling waves, center, break over long distances, gradually losing their energy. Surging waves appear as "humps" moving through the water and do not even appear to break.

Figure 7. The two most common types of breaking waves, plunging breaker (top), and spilling breaker.

ing or surging waves. Shorebreaks occur when large waves break directly on the beach.

Spilling breakers are much less imposing and lose their energy over long distances. They are formed when waves move over beaches with gradually sloping bottoms. The breaking water rolls or tumbles forward as the wave advances into shallower water, producing a wide surf zone. Spilling waves generally provide safe conditions for waders, inexperienced swimmers, and novice boogie boarders; the East and Gulf coast beaches are most often subject to this type of breaking wave.

Surging waves are much more rare than spilling and plunging waves. These waves are created where the water is relatively deep near shore cliffs and coral reefs or at very steep beaches composed of gravel or small stones (called shingle in Great Britain). Surging waves can be deceptive because they do not truly curl or break; instead, the surge causes a sudden rise and fall of the water level. Serious injuries have been caused by surging waves on rocky coastlines.

What Causes Rip Currents

Rip currents are caused by large amounts of water being pushed far up the beach by breaking waves. The water then escapes back to the ocean as concentrated flows. Plunging breakers of large swell waves are the most effective in producing the conditions for rip currents. There are exceptions, however: waves generated by strong local winds blowing directly onshore are responsible for most rips along South Florida beaches, as the summer swell waves are effectively blocked by the Bahamas.

When a wave breaks, it produces swash, which is the water that moves up and down the beach face. Bigger waves contain more water, and plunging breakers produce the most white-water, propelling the swash far up the beach. This water that is forced onto the beach, then naturally flows back down the slope to the sea surface. Water will follow the path of least resistance, such as an underwater trough or a protected area beside a groin, in seeking its own level. A concentrated flow of water returning to the ocean becomes a rip current (figure 8). The bigger the breaking waves, the more water that is trapped on the beach face, and consequently the stronger the rip. Therefore, rip currents vary greatly in size and power. "Rip tide" is a misnomer because tides play little to no role in causing these strong offshore-flowing currents (although along certain beaches, rip currents are strongest near the time of low tide).

Waves contain the energy that generates nearshore currents. These currents are the ones that primarily affect bathers and swimmers and extend from the shoreline to the outermost breakers (the surf zone). Tidal jets are another dangerous current (totally unrelated to waves) that occur at inlets or other constrictions; these strong currents are caused by the flooding and ebbing tides.

Some beach communities, especially in Southern California and Hawaii, post signs warning swimmers of undertow during big wave days. Undertow is the one thing that many beachgoers have heard about, yet it does not really exist. It supposedly pulls you under the water, but it is actually the swash backwash or rip current, which pulls you seaward at all depths. The sheet of water moving up the beach from the broken wave

Figure 8. Rip current at Ocean Beach, San Francisco.

is called the swash uprush. The swash is then drawn by gravity down the beach (backwash) as water seeks its own level. When the waves are high and the beach is steep, the swash backwash can be deep and powerful.

On big wave days, especially with large plunging breakers, the swash can be very strong; the water shoots up the beach face, providing a good ride for boogie boarders. But the backwash of the swash is problematic on steeply inclined beaches near the time of high tide, especially for women and children, who generally have a higher center of gravity. This return flow of water can topple people; it is difficult to maintain your footing in the swift current as you are pulled forcefully toward deeper water. While this current can be overpowering during times of big plunging waves, it will not take you beyond the breaker line (unlike a rip current, which carries you offshore through the surf zone). Of course, it can be dangerous if you are pulled into the next large, plunging wave that is breaking into shallow water.

The most frequently encountered current by bathers and swimmers on ocean beaches is the longshore current, which is produced by waves breaking at an angle to the shoreline. Anyone who has spent time on surf beaches has experienced this current—it moves you along the shore, but not offshore. Sometimes the current is so gentle that you don't even feel it moving you; it is not until you get out of the water and can't find your towel that you realize its effect. Longshore currents can be hazardous to weak or non-swimmers when the current moves them into a hole where the water can be above their heads. At this point they can panic and drown. Other times, especially when the breaking waves are coming from an oblique angle to the shoreline and are large, this current can feel like a river flow (although you should not even be in the water during these conditions). In fact, the longshore current is responsible for moving huge quantities of sand.

Recognizing and Escaping Rips

The United States Lifesaving Association (USLA) estimates that rip currents are to blame for 80 percent of all ocean rescues. This statistic indicates how important it is for people to recognize this hidden danger. While lifeguards and frequent beachgoers can detect such currents readily, those unaccustomed to surf beaches may have difficulty spotting them. Waves break in a water depth that is 1.3 times their height. Because rips flow through underwater channels or holes in the sandbar, waves do not break as readily there. Also, the force of the rip current itself tends to diminish the power of the incoming waves, lowering the surf. As already shown with the family visiting Ocean City (see Preface), some unsuspecting bathers and swimmers can even be attracted to rips because of the calmness of the water relative to the high surf elsewhere.

People must learn how to spot rips because it can be a matter of life and death; even expert swimmers can be nearly helpless in these powerful currents. Rips are sometimes referred to as the drowning machine because of their almost mechanical ability to tire swimmers, ultimately causing death. The first thing I do on arriving at a beach is to scan the surf from the highest point possible. Then I check for warning flags and consult with lifeguards on surf conditions, especially regarding the presence and location of any rip currents. Rips do not always appear in the same spot—they can change position from day to day and week to week. More than one rip may be present at the same beach on the same day. I always look for a seaward flow of debris such as seaweed and foam or suspended sediment. In rip currents, these materials generally move at right angles to the shoreline. Where the rip crosses the surf zone, the line of breakers may be interrupted or transformed into small, choppy waves. Also, the water contained in the rip often looks murky or foamy. Rip currents moving through the relatively calm, regular surf of big Pacific swells, such as along the Southern California coast, are easily detected. These deadly currents are much harder to spot when the sea is rough and conditions are windy.

Rips have three components: feeder, neck, and head (figure 9). The feeder current, which is often a longshore current, is the main source of water for the rip. Water that has been pushed and piled up on the beach is often moved along the shore for a short distance by the feeder currents to the underwater channel or trough. Once the water reaches the channel or encounters an obstacle to its along-the-shore movement, it will turn seaward as a rip current. There may be one or two feeder currents, depending on the wave approach and prevailing longshore current.

The neck is where the concentrated flow of water moves from the beach through the surf zone. Current speeds frequently

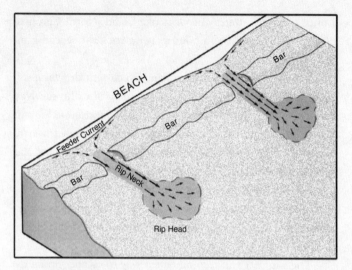

Figure 9. Rips generally flow seaward through breaks or low points in the bar system.

reach two to three feet per second (the speed of a champion swimmer); speeds as high as five feet per second have been recorded along some Australian high-surf beaches. The neck of the rip can vary in width from a few yards to tens of yards. The majority of both rescues and drownings happen when people are being pulled offshore in the rip neck.

The rip head, which has a mushroom shape, develops where the current has moved beyond the surf zone. Here the rip loses its power as the water disperses broadly. There is no longer a current, and anyone being pulled to this offshore area can then just swim back to shore, avoiding the narrow rip neck.

I once tried this strategy at Ocracoke Island, North Carolina, but it took me almost an hour to swim back the several hundred yards in high seas.

In a common scenario involving rip current drownings, it is the rescuer who drowns and the initial victim who survives. Frequently, a child who is caught in a rip will scream to his parents, who rush to his aid. But without knowing the best way to escape the rip current, the parent often drowns. To avoid this sort of disaster, always swim at a guarded beach. If you attempt to rescue someone, take along a flotation device. A boogie board or surfboard is ideal, but a Styrofoam cooler will often be buoyant enough. If a rope is available, stay onshore and throw it to the victim.

The simplest and recommended way to escape from a rip is to swim perpendicular to the pull of the current (that is, parallel to the beach). Unfortunately, most people caught in rips tend to panic. Our survival instinct tells us to swim back toward the beach, which is absolutely the wrong thing to do. Even experienced swimmers who attempt to swim against strong rips will fatigue and eventually drown.

East Coast

My first beach experiences were at Myrtle Beach, South Carolina, when it was still a small beach town, consisting mostly of one- and two-story white wooden houses. As a boy, I lobbied my parents to return to the beach every summer as opposed to making a mountain trip, because playing in the waves was so

How to spot a rip

Look for signs of rips before entering the ocean

• Change in water color from the surrounding water (either murkier from sediments, seaweed, and flotsam, or darker because of the depth of the underwater channel where the rip flows).

• Gap in the breaking waves, where the rip is forcing its way seaward through the surf zone.

• Agitated (choppy) surface that extends beyond the breaker zone.

• Floating objects moving steadily seaward.

• Temperature of water—water in the rip may be colder than the surrounding water.

What to do if caught in a rip current

• Don't panic, which wastes your energy and keeps you from thinking clearly.

• Don't attempt to swim against the current directly back to shore.

• Swim parallel to shore until you are out of the current as the offshore flow is restricted to the narrow rip neck.

• Float calmly out with the rip if you cannot break out by swimming perpendicular to the current. When it subsides, just beyond the surf zone, swim diagonally back to shore.

• Wave your arms in the direction of the lifeguards.

much fun. I was a good swimmer and had no fear of the water; only my parents' waving kept me from going too far offshore. The waves were relatively small and safe but still fun for a kid to ride. Rips rarely occur at Myrtle Beach, which was fortunate because I had no knowledge of them. The sand there is very fine, which makes the beach flat and slope gently into the water. Barring a hurricane, it is a very safe beach for families.

Rips are not present in flat water, such as ponds or bays, and are not a problem when wave action is limited. The tides on the Georgia coast are large, but the waves are small (don't bother packing the surfboard), and actually, the two are related. The shallow continental shelf extends out for hundreds of miles, which contributes to the larger tidal range along with the shoreline configuration (the "Georgia bight," which is a concavity amply clear on maps of the East Coast). The rising water is simply "bunched up" so that the tide must come in higher along the Georgia and southern South Carolina coast relative to the states on the flanks—North Carolina and Florida. The larger waves break many miles offshore.

The first time I saw a rip current was at the groin field near Cape Hatteras, North Carolina. Many people confuse groins and jetties. Groins are coastal engineering structures, typically made of wood, sheet metal, or rock, that extend into the water at right angles to the beach. Groins are often built in a series along the beach, constituting a field. By contrast, jetties are used to stabilize inlets for navigational purposes and are much larger in size, constructed out of large boulders or concrete forms (tetrapods).

The groins at Cape Hatteras were wooden and fairly short,

but rip currents are commonly produced on this high-energy beach as the strong longshore current is directed offshore by these blocking structures. Experienced surfers routinely use the rips to take a free ride offshore on their boards, and then ride the breaking waves ashore. Swimmers should avoid being near groins in all cases because of the possibility of being slammed into one by breaking waves or being swept offshore by rip currents.

The Outer Banks of North Carolina are famous for many things—the first English colony in the New World, the place where the Wright Brothers first took flight, and the home of Blackbeard the pirate. In recent years it has become a surfing mecca because of the large swell-type waves that arrive, especially in late summer. Duck, a community named for the hunting camps of an earlier time, has become one of the favorite beach escapes for Washingtonians and other urban dwellers looking for upscale restaurants and accommodations. I like to gather together family and friends and rent one of the large wooden beach houses on the high rolling sand dunes that characterize this section of the barrier island. The beach is composed of coarse sand and reddish pea-sized gravel. The coarse sand is indicative of the presence of a beach steeply grading into deeper water so that waves can come close to shore before breaking. I recommend that families with children visit these shores early in the summer, when the waves are relatively small. During the late summer, especially August and September, the Atlantic Ocean becomes stoked by tropical storms and hurricanes in the tropical latitudes. Big swell waves can travel for thousands of miles without losing their energy until they break on the beach. In these cases, dangerous surf can be produced, as well as life-threatening rip currents. A number of people have drowned on the Outer Banks in recent years because they do not understand or recognize these powerful currents.

The most instrumented beach (equipped with current meters, wave gauges, bottom-sounding equipment, and other devices) in the world is at Duck—the U.S. Army Corps of Engineers Field Research Facility. Touring this facility (a bombing range during World War II), you will discover how coastal scientists and engineers make measurements of the waves and currents, as well as record changes in the beach and nearshore areas. A high-resolution videocamera has been installed in the tower overlooking the beach. This "beachcam" provides streaming video of the surf zone, which can be used to analyze the changing position of the underwater sandbars (by inference of the breaking wave patterns) and possible presence of rip currents.

Moving through underwater channels or breaks in the inner bars, rips at Duck appear to persist for weeks or even months at the same places. Only a particularly large winter storm like a nor'easter or a hurricane moves these channels—conduits for the seaward-flowing currents. Some rip currents have lasted from May until September, according to Robert Dean, a colleague of mine at the University of Florida. This finding contradicts the prevailing view of these currents as spontaneous and short-lived. While they may exist at the same position for months, they become strong (and hence noticeable and dangerous) only with the approach of large swell waves producing plunging breakers. Certain sections of the shoreline appear to be more prone to rip currents than others, but no systematic studies have ever been undertaken.

Ocracoke Island, south of Cape Hatteras, is accessible only by ferry or small airplane. I enjoy visiting this island, which was home to Blackbeard before he was killed here and beheaded. Those were wilder times, but the 16-mile long beach at Ocracoke can be pretty wild itself with big, rolling breakers. In earlier times, I was in peak shape and a good swimmer. Once, a few buddies of mine decided to personally test the strength of a rip current, not the brightest idea that we ever had. As it turns out, the rip was bigger than we had anticipated, and we were pulled a few hundred yards offshore into fairly deep water before being released from the current's grip. I had not counted on the difficulty of swimming ashore in breaking waves that approached ten feet in height. Wave size is limited by the depth of water; waves can reach a maximum height of 0.6 times the water depth. I always tell my students that you cannot have five-foot waves in a mud hole. In this case, we were well beyond normal limits. It took us nearly an hour to swim back ashore, and we were exhausted from this experience—one I do not intend to repeat.

I have spent many pleasurable summers working on the south shore of Long Island, New York. This area gets some good wave action, especially during late summer because of the tropical storms in the lower latitudes. Jones Beach State Park on Long Island's south shore is the most popular and heavily visited beach on the northeast coast. Over 6 million people congregate each year on this strand of barrier beach. The beach resembles a patchwork quilt from the vantage point of my overflights—beach towels nearly touch end-to-end and side-to-side. Surprisingly, the people who frequent Jones Beach like it that way. This is truly a touchy-feely beach, but it will never rank high in my ratings, as the most popular beaches are rarely the best.

The water at Jones Beach is chock-full of waders and swimmers on summer weekends. Strong rips occasionally develop, and the mere multitude of people forces unwitting bathers into the swift offshore-flowing current. Lifeguards have had to pull out hundreds of people from a single rip current during an exhausting weekend of duty. In this special case, the presence of the rip is recognized, but the "crowd pressure," like rush hour on the subway, can nudge you into the current.

Further west along the south shore of Long Island at the Rockaways, fierce rips have killed many swimmers through the years. During July 2001, three teenage girls were wading in knee-deep water at Far Rockaway in Queens when they were swept away by a rip toward Reynolds Channel (a tidal inlet with a strong current) and then offshore to their deaths. One of their mothers stood on the shore helplessly and said that the ocean "just opened its mouth and swallowed them up."[1] This area has been called a death trap, and few dare to swim here, even with lifeguards on duty. It is advisable to avoid swimming and wading near an inlet, as the tidal currents can carry you far offshore. This area seems to be a double whammy (rips and tidal currents) for beachgoers who want to take a dip.

Rips typically form in pronounced breaks or "holes" in the nearshore sandbars that serve as the conduits for the strong seaward-flowing current. Such strong currents could scour holes in the inner bar, but what caused the rip to form in the first place? This is the proverbial chicken-and-egg question. These holes in the inner bar also cause problems of exacerbated beach erosion during storms, as the waves break closer onshore and with more energy in these areas. Localized beach erosion at Fire Island and the Hamptons, the barrier beaches to the east, occurs during storms because of this phenomenon.

Most of the drownings in the Hamptons occur in August, and the summer of 2001 was no different. A 23-year-old man was caught by a rip current and pulled away from shore into deep water, where he drowned. Blue flags are flown by the towns of Southampton and East Hampton to indicate strong currents, but not everyone heeds these warnings or swims at beaches protected with lifeguards. August is the time when the water is the warmest, when the most people are in the water, and when hurricanes hundreds of miles away are kicking up the waves that come ashore here as swell-type plunging breakers. Clearly, people need to recognize the telltale signs of rips to avoid a similar fate.

I spent many summers on Cape Cod while directing the National Park Service Research Unit at the University of Massachusetts in Amherst. One of the beaches of particular interest was Nauset Spit in Chatham. There are no paved roads on the long barrier beach, so access is by boat across the bay or by off-road vehicles. In the late 1970s, while heading up an Earthwatch expedition, I stayed in the Old Harbor Station—one of the original late 1800s Life-Saving Stations, complete with tower. It was a great experience living in this landmark building, which had been brought close to the water's edge by long-term erosion. At high tide the water nearly lapped onto the front porch. At low tide, the beach was much more expansive, as the tides here are very large (six feet and more during full moon). The nearshore bar is nearly exposed at dead low tide and often exhibits a crescent shape with low areas or "holes" in the sandbar at a regular spacing of several hundred feet. During mid- to high tide with good surf (four- to five-foot plunging breakers), I have often observed rip currents being set up, with this swift current moving through the holes in the bar. I advised my team members not to go swimming during these conditions in the chilly water (it rarely reached 65 degrees Fahrenheit, even on a hot summer day). While we were conducting some research elsewhere on Nauset Spit, a teenage boy went swimming and was swept offshore by a rip current right off the Old Harbor Station. We all were saddened to learn of his death, and I wish I had been there to warn him of these silent killers. Rips are weak or not present during low tide because there is not enough water depth. From mid-tide to high tide, the inner bars trap the wave runup (swash) of the plunging breakers; this elevated water then flows alongshore to the holes in the bars, where the rip currents form.

I often take my two children to the surfing spot at the point on South Beach, Miami Beach. They like to boogie board but must watch for the long-boarders who try to surf right into the swash zone. One afternoon the lifeguards were flying red flags for a rip current alert. There were many children in the water, and the waves were low—only a few feet high and obviously not dangerous. I scanned the clear, greenish-colored water and saw that there were no rips present. I asked the young lifeguard about the warnings. He said that there had been rips in the morning and was disinterested in further conversation. While I give the Miami Beach lifeguards high marks overall, this was an unfortunate incident. Such false warnings give people the wrong impression about rip currents in particular and beach safety in general. The next time rips are actually lurking, people may not take the lifeguards' warnings seriously.

The surf in South Florida is generally low during the summer, but the winter nor'easters that cause considerable erosion along the mid-Atlantic and northeastern coasts produce some big swell waves in these subtropical areas. The strongest rips often occur during the winter and early spring, when the long-crested swell waves approach the beach directly onshore. In fact, many rip rescues occur during spring break because of the large number of visiting college students with limited experience swimming in the ocean. While the swell waves are generated far out at sea, strong, local onshore winds can also generate large surf. Rips are expected on days with such winds and become particularly dangerous as the breaking waves approach four to five feet in height. Statistics indicate that most of the drownings associated with rips occur at low tide on this microtidal (less than three feet) shore. This is because at low tide there is more whitewater (e.g., more powerful swash) produced by waves breaking in the shallow water over the bars, and the piled-up swash has to escape through the breaks in the bars as rips.

The vast majority of rip current rescues in Florida occur in Volusia County, on the state's northeast coast. In a typical year thousands of people are pulled out of these murky waters, especially during the summer. (In these same waters, especially at

New Smyrna Beach, shark bites are frequent.) The large number of rip incidents there may be related to the popularity of the beach and to its long length—it extends tens of miles.

Gulf Coast

The Gulf of Mexico is not the ocean, which is apparent to anyone accustomed to the pounding California surf or the summer swell-type waves of the Atlantic Ocean. This is good news for people prone to seasickness like myself; an offshore boat trip in the Gulf means cutting through small chop rather than pitching and rolling in heavy seas.

The Gulf Coast is characterized by beaches with long, gradually sloping bottoms, which generally prevent the build-up of pounding surf on the beach. The Gulf waters are usually placid to choppy, so that big breaking waves occur only when there is an offshore storm in this great water basin—that is, the surf is up only when tropical storms come knocking. This means that rip currents are infrequent here relative to Atlantic and Pacific beaches. When the water is clear and the surface calm, submerged crescent-shaped sandbars may be visible in the nearshore area, close to the beach face. The presence of this unique "rhythmic" topography of the inner bars sets the stage for rip current formation if large breaking waves approach the beach.

Panama City Beach is located in the Florida panhandle along the famed Emerald Coast, so named because of its clear, turquoise waters. This long strand of barrier islands also boasts the finest, whitest sand in the world (although someone who has explored a beach in Kenya is now challenging this claim to fame). I can remember the first time I ever saw these sugar-sand beaches as a young boy. I started running into the warm waters, expecting to tumble forward when tripped by the deepening water. Instead I just kept running and running and still the water was not that deep, much to my surprise. The superfine sand makes for a very flat beach that gently slopes into the water, an exceptionally safe beach in combination with the almost placid seas. But any beach can be dangerous on a given day.

Panama City Beach is where Choule Sonu conducted pioneering research on rip currents. He chose this area of Bay County for his field experimentation because of its clear, warm waters. Fluorescent green dye was used to trace the motion and to time the speed of the currents. During five-foot waves, which rarely break on these normally placid shores, Sonu measured maximum rip speeds at two to three feet per second. The nearshore bar often contains large gaps in which rips can naturally form when the seas are pumped up by an offshore storm. Sadly, Bay County has experienced 24 rip current drownings since 1989; many of the victims are from the heartland of America and are not familiar with the ocean.[2]

Pacific Coast

Surfing in Southern California was immortalized by songs such as "Surfin' USA" by the Beach Boys and "Surf City" by Jan and Dean. The big ocean swells that rhythmically beat along these shores make for great surfing waves. The plunging breakers also generate large swash that "sets up" (elevates) the water on the beach, moving it well above the average water level. Some of this elevated water moves back seaward as backwash, but a portion is often moved along the shore by feeder currents to a rip. I have often marveled at this well-organized nearshore circulation system, which regularly spaces large rip currents along the shore. The tell-tale plume of seaward-flowing water caused by these big rips is apparent from the beach, but it is easiest to spot from cliffs or fishing piers. I always take time to gaze at the impressive rip currents along Huntington Beach, California, when visiting my relatives. The community of Huntington Beach is known as Surf City, USA, and surfers are not bothered by the rips; indeed, they use them to take a free ride offshore to catch the next set of big waves.

Rip currents are the bane of Southern California beaches for swimmers. Tens of thousands of people are rescued each year by the region's well-trained lifeguards. Although most of the action at Venice Beach is on the concrete boardwalk, the lifeguards who roam the beach in their bright yellow, four-wheel-drive trucks make many rescues.

Ocean Beach in San Francisco is where I take television reporters looking to photograph rip currents. The best way to see the full extent of these offshore-flowing waters is from the air. For the television shoots, the producers rent a helicopter for us, and I drop fluorescent orange dye to visually trace these powerful currents. Rip currents are a regular occurrence along Ocean Beach, the only ocean beach in San Francisco, and swimming is never permitted there. Unfortunately, each year some out-oftowners try to take a dip and are pulled several hundred yards offshore, where they drown in the frigid waters. If Ocean Beach were renamed Killer Rip Beach, perhaps visitors would then take note and stay out of the water.

Anyone unfortunate enough to be sucked offshore by the big rips at Ocean Beach would also be subject to the strong tidal currents. During a flooding, or rising, tide, your body, long since chilled by the very cold waters, would be carried beneath the Golden Gate Bridge and past Alcatraz Island. It has been said that no one ever escaped Alcatraz when it was a maximum security prison because of the strong tidal currents and hypothermia that swimmers experience within half an hour.

The most popular swimming and surfing beach in the San Francisco area is Stinson Beach in Marin County. Stinson has good amenities, and the beach is very popular, particularly with those who can tolerate the water, which at its warmest is still chilly by East Coast standards. When the big Pacific swell starts coming ashore, one and occasionally two big rips are set up near the middle of this pocket beach. The lifeguards mark the rips with flags to warn swimmers. At Stinson Beach you can have it all—big powerful rip currents and the most fearsome of all sharks, the great whites. Although this beach definitely sounds like a place to avoid, there are remarkably few drownings, and shark attacks are generally just bites to the legs and feet of surfers, although some people have been killed there.

Hawaii, America's tropical paradise, boasts some of America's best beaches. Many of my national winners are Hawaiian beaches, and I have often been asked why all the top beaches are not in these tropical isles, where the mountains meet the sea. The major reason involves beach safety because of the big waves, shorebreaks, and dangerous rip currents. What is interesting about Hawaiian beaches, where the water is warm enough to go swimming year round, is how much the waves change seasonally.

The big surfing beaches are on Oahu's north shore, the surfing capital of the world. Here the Pacific Ocean swells normally reach 20 feet in height and can tower to an incredible 40 feet. Waimea Bay is famous for having the largest ridable surfing waves in the world. This U-shaped pocket beach is bounded by cliffs, which offer great views of these terrific and terrifying breakers. The plunging breakers form a tube of air through which brave (and some say foolish) surfers take the ride of their lives. During winter, waves breaking on the offshore coral reef provide some of the most awesome and dangerous surf imaginable. Lifeguards use jet skis for rescues, but they must enter the mountainous seas at just the right time. During summer, these same waters can be as quiet as a lagoon. The contrast is startling, and it makes you wonder if you are really at the same place. While swimming there is a pleasure in the summer, don't even think about getting into the water from September to April.

Most rip currents are generated by large waves, usually plunging breakers. But other conditions can also generate rips. All that is really required is to set up the water above sea level and have a constricted area for the return flow. Such a location exists at Kee Beach, at the literal end of the road and gateway to the Na Pali Coast, Hawaii's premier wilderness experience.

Kee Beach, on the island of Kauai, has sensational scenery. Sheer mountain cliffs with intermittent waterfalls provide a dramatic backdrop for this small pocket beach. The quiet blue-green water serves as one of Hawaii's most popular snorkeling sites, where you can swim with schools of brightly colored fish. Waves breaking over the coral reef, exposed at mid- to low tide, result in elevated water being trapped at Kee Beach. This super-elevated water cannot escape easily through the coral rock that surrounds this little beach, except through a channel on the western end of the reef (figure 10). Unsuspecting swimmers and snorkelers paddling around the quiet waters of this lagoon-like setting can suddenly find themselves caught in a powerful rip current, which has caused a number of drownings. The lack of high surf at the water's edge does not always assure safety, so lifeguards should always be consulted when entering the water for the first time.

Tales from Abroad

Australian beaches, which are some of the most beautiful in the world, are known for their great white sharks and killer rip currents (but not necessarily at the same beach). There is even a beach in Western Australia named Shark Beach, but the name relates to earlier explorers catching and eating sharks here, not the other way around. Rips in the clear tropical water on the big wave beaches are often regularly spaced along the shore. Because of the great concern for beach safety, largely due to the powerful rip currents, Andy Short of the University of Sydney

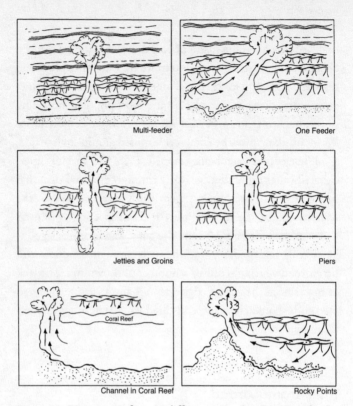

Figure 10. Rip currents form in different ways. *Source:* U.S. Life-saving Association.

has made a career studying their occurrence and distribution along the hundreds of beaches that surround this island continent. Short rates beaches according to their dangers and risks to beachgoers; I identify these problems as part of my 50 variables in combination with all the positive aspects in my annual rating of America's Best Beaches (see appendix E).

A good friend and colleague, Allan T. Williams of Wales, had a harrowing experience with rip currents. During his doctoral research at the University of Hong Kong, he was working at Repulse Bay while a small typhoon (hurricane in our parlance) was making landfall. As a good coastal geomorphologist, he carefully surveyed the beach beforehand so that he could compare it with post-storm measurements to determine the amount of erosion. His beach profiles extended underwater to a depth of 30 feet, which required scuba. With the formal survey completed, he watched the waves continue to build up as the typhoon came ashore. Williams thought that it would be interesting to see how the waves were moving the sand in these clear tropical waters, so he suited up again and made his way into the surf between large breaking waves. Soon after entering the water he felt the tug of a rip that was much more forceful than any other current he had ever experienced. He was quickly pulled offshore. About 400 yards out into the harbor was a large channel buoy, moored to the bottom by a metal chain. Somehow he managed to grab hold of the chain and pull himself up onto the platform of the buoy, where he rode out the rest of the storm. Luck was clearly on his side.

three**Beach Smart**

During spring 2001, I made a trip to Thailand to evaluate possible environmental impacts from producing the Hollywood film *The Beach* on location. After poring through all the technical material and experiencing this sensational beach myself, I took a few days off to snorkel in the tropical waters. While swimming amid a school of multicolored fish, I spied what at first appeared to be a giant sea worm wiggling through the water beneath me. Then it dawned on me that there were no giant sea worms in the ocean—this was a poisonous sea snake! The adrenaline rush was almost intoxicating; I found myself wanting to linger for a long look, but common sense took over, and I swam away. I didn't want to be in the way when the snake came up for a breath of air. Sea snakes have several different kinds of deadly poison, and their bite is almost always fatal. In actuality, these creatures are generally docile, and fishermen pulling nets aboard their boats are the ones most frequently bitten by the entangled, and hence angry, sea snakes.

Any beach can be dangerous on any given day. The ocean is not a swimming pool, and waves make a beach exciting but occasionally dangerous. But there are many other hazards at the seacoast to contend with as well: overexposure to the sun, lightning strikes, water pollution, and stinging animals, among others.

Ironically, the most common type of injury at seashores is splinters from wooden boardwalks. I wear flip-flops; they are easily removed and light enough to carry when walking the beach. (I consider it almost a sin to walk on good-quality, sandy beaches with shoes.) Splinters are more an irritation than a hazard; the biggest threat to beachgoers is sunburn.

Sunburn and Skin Cancer

The human craving for sun is natural, and there is no better place to enjoy both the sun and the surf than an ocean beach. There is no doubt that the sun lifts our mood and even affects our state of mind, but we need to be careful because too much of a good thing can be harmful.

Sunburn tops the list of beach perils; the red lobster look is both painful and unhealthful. Baking in the sun can cause considerable skin damage and eventually lead to skin cancer. Children are most at risk. Even one bad sunburn can, later in life, lead to the serious and sometimes fatal skin cancer called melanoma. Fair-skinned people are especially vulnerable to this underrated killer.

Sunburn is a particular problem at beaches because of the sun's strong reflection off the beach and water, which can greatly amplify the amount of solar radiation your body receives. Most experts recommend using products with a sun protection factor (SPF) of 25 or more. Remember to apply sunscreen lotion on your feet, which burn easily.

The sun is much more intense in lower latitudes like Florida and Hawaii, and special care should be taken during the summer. To protect yourself, use sunscreens and avoid the peak hours of ultra-violet (UV) radiation during "solar noon" at 1 P.M. when the sun is the highest in the sky. I usually leave the beach by 11 A.M., have a leisurely lunch, read a book or magazine, and then return to my open-air activities around 3 P.M.

The incidence of skin cancer is rising, affecting more than 1 million Americans a year. Some 50,000 people are diagnosed annually with melanoma, which represents a staggering 1,800

Skin cancer warning signs

- Asymmetrical moles

- New moles appearing on adults

- The appearance of scaly or crusty areas on the skin

- Moles with multishaded areas

- A mole that begins itching and/or bleeding

- Existing moles that enlarge irregularly

Source: American Cancer Society (www.cancer.org). See also www.epa/gov.sunwise.

percent increase since the 1930s. Australians are even more affected because of the thinning of the protective ozone layer,
which permits the transmission of stronger UV radiation to the
earth's surface. Melanoma, the least common but most deadly
form of the disease, is a unique kind of cancer in that it "writes"
its signs on the skin. If caught early, it is highly curable. Be
beach smart—don't get too much sun, use sunscreens liberally
and often, and recognize and heed the early warning signs of
skin cancer to stay alive. Skin cancer is something that people
in my profession really have to worry about because we spend
so much time in the sun. Besides, I do not want my skin to look
like my name—Leatherman.

Lightning Strikes

Lightning strikes are a much-underrated beach hazard. The
largest number of people killed by lightning are either standing under trees, such as at golf courses, or exposed on beaches.
Some people use umbrellas to block the sudden downpour of
rain during a summer thunderstorm, but an umbrella can serve
as a lightning rod, with devastating results.

It is estimated that 25 million lightning bolts hit the
ground each year in the United States—these are nature's fire-
works heating the air to an incredible 50,000 degrees Fahrenheit! Lightning is the second leading cause of weather-related
deaths, following floods. On average, 73 people are killed and
300 are injured by lightning strikes each year in this country,
reports the National Weather Service. Florida is the lightning

capital of the United States, with about ten deaths a year—almost twice as many as any other state. The greatest danger often comes from the first or last flash of a storm, when people may not be under cover. This hazard is all too often ignored.

In June 2001 a photojournalist vacationing in South Florida was standing on Fort Lauderdale Beach. It wasn't raining, but according to his companion, thunder could be heard from a storm that was several miles away. Suddenly a lightning bolt hit the man in the head, killing him instantly and injuring his companion and six others nearby. This story illustrates one of the little appreciated facts about lightning—namely, it doesn't have to be raining in order for lightning to strike. In fact, more people are struck before it starts raining or after it stops than are struck during the height of the lightning storm.

Lightning can strike from ten miles away, so if you can hear thunder or see lightning, you are in danger. Take shelter indoors or in a hard-topped vehicle. Even though it is tempting to rush back onto the sand after the rain ends, the safest thing to do is wait at least 5 and preferably 30 minutes after the last thunder is heard before going back outside.

Other popular beach activities such as wading, swimming, snorkeling, scuba diving, and jet skiing are hazardous when lightning is near. Water, especially salty or brackish, is a good conductor of electricity. If lightning strikes the water, it can electrify the surrounding area. As an example, a few years ago two boys in Coconut Grove, Florida, were swimming in a community pool. Lightning hit the water, knocking them both unconscious and causing them to drown.

Beach Lightning Safety Rules

• Listen for weather reports before going to the beach to see if lightning storms are forecast.

• While at the beach, watch for approaching dark clouds, especially during the afternoon coming from an inland direction. Since most people lying on the beach are facing the water, they may not notice the buildup of thunderstorms behind them. A radio can act as a lightning warning device—listen for static.

• If thunder is heard or lightning is seen, immediately get out of the water and off of the beach. Seek safety in your hard-topped vehicle or inside a fully enclosed structure. Cabanas or tiki huts without sides are not safe places.

• Avoid being near trees or other tall objects such as antennae.

• Wait at least 5 minutes (and preferable 30 minutes) after the last thunder is heard before returning to the beach.

• If someone is struck by lightning, immediate first-aid may save his or her life. Lightning frequently causes the heart to stop, but CPR can restart it. People struck by lightning carry no electrical charge.

Source: Lightning Safety Website: www.lightningsafety.noaa.gov

Structures on the beach, such as lifeguard towers or cabanas, can often be the target of lightning. It is prudent to have lightning rods on all such structures. Trees, especially solitary tall ones, such as palm trees, are dangerous to be near. Lightning can hit the tree and then flash off to strike people standing nearby. If taking shelter in a lavatory, avoid touching plumbing fixtures.

First Flush

The term first flush is used to describe the effluent coming off a parking lot during the first rain in the period of a week; stormwater drains on beaches present a problem for some urban areas. The Surfrider Foundation has been fighting such water pollution problems for years in Southern California, with considerable success. The Natural Resources Defense Council maintains records of beach closures and annually issues an on-line report, "Testing the Waters" (www.nrdc.org).

There are thousands of beach closures each year caused by high pollution levels. The biggest problem is storm drain runoff from lawn fertilizers, pesticides, dog and bird feces on streets and sidewalks, and illegal dumping. In the worst cases, thunderstorms that produce copious amounts of water result in sewage system overflows of untreated effluent that are released directly into the ocean. Typical symptoms of swimming with pathogens in the seawater are exotic skin rashes, sinus infections, and raw throats.

North Myrtle Beach, South Carolina, decided to take an

How to protect yourself in polluted waters

• Wait several days after a rainstorm to swim or surf in places where possible runoff contamination is high.

• Swim at least 100 yards from storm drains that flow into the ocean. Note direction of the longshore current and swim updrift.

• Avoid swimming at beach areas around marinas, where bacterial levels are often high.

• Never allow your children to play in ponded runoff or storm drain water on the beach.

engineering approach to deal with water quality problems. The city recently constructed two 1,300-foot-long pipelines to shunt the bacteria-laden runoff far offshore. This allows the salt water to kill the freshwater bacteria at a safe distance from the beach. Past problems with water quality occurred during the summer after heavy rainfalls, causing as many as 12 health warnings during the swimming season. Storm runoff is no longer a problem for this area; the beaches are now both clean and safe.

The United States is blessed with many great beaches with water quality that is generally quite high. Overall, our beaches are much cleaner than most found elsewhere in the world, but problems remain and more needs to be done to maintain good bathing conditions.

Red Tides

Red tide, which occurs when there is an algal bloom, is a serious water quality problem. You can spot a red tide (which can
also be green or brown) by masses of floating organic matter on
the water surface. In the worst cases, these floating patches can
stretch for miles. The appearance of an algal bloom means that
people should be cautious at affected beaches. The toxins most
often cause illness in humans who swim in affected waters or
eat contaminated oysters. Even toxins carried on the sea breeze
can cause flu-like illness. These toxic algae (dinoflagellates), re-
producing by the billions, can cause massive fish kills. Infected
shellfish are poisonous to eat.

The Gulf coast is impacted more often by dreaded red tides
than other U.S. coasts. Most problems occur during spring to
early summer, resulting in murky water and piles of algae and
dead fish washing up on the beach. While red tides are a natural phenomena dating back at least to Biblical times, some marine scientists believe that their apparent increased frequency
is caused by water pollution, specifically excessive nutrients
(like phosphates and nitrates) running off the land. Fortunately
for beachgoers, the really toxic alga called the "cell from hell"
(Pfiesteria) has been found only in estuaries and bays, not on
ocean beaches.

Jellyfish Alert

Jellyfish are a bane to swimmers, and the Portuguese man-of-war
is the worst of the lot in terms of stinging potential (although

technically it is a hydroid and not a jellyfish). These iridescent, purple-colored animals have floating bubbles and long, stinging tentacles ranging up to 50 feet in length. One inch of a tentacle contains 900 stinging cells that fire on impact, injecting toxin into the victim. This acidic venom can cause great pain, shock, and, in rare cases, even death. Usually these jellyfish are blown ashore by high winds, particularly during stormy conditions (similar to those that cause rip currents). The sting can be neutralized by applying full-strength vinegar, rubbing alcohol, diluted chlorine bleach, or plain meat tenderizer, which should be applied immediately to the wound. Victims should be kept out of the sun until the pain subsides. People with allergic reactions should be rushed to a hospital. A few hundred thousand people are stung by Portuguese men-of-war each year along the U.S. East Coast—what a beach bummer!

Other marine animals that should be dealt with cautiously include sea urchins and stingrays. Urchins are found worldwide and are one of the more common marine-life hazards. They are covered with sharp, needlelike spines that can penetrate deeply into your foot if stepped on. Sea urchins are found attached to rocks or coral. By contrast, stingrays sometimes come into shallow waters and lie on the sandy bottom, where they are nearly invisible. The calm, shallow waters just off the beach in the Gulf are inviting to stingrays. Shuffling your feet on entering the water will scare away these shy animals. They often move closer to shore in August; it is best to consult a lifeguard, who might have noticed any major influx of stingrays. Stingrays constitute a much more frequent danger than sharks to bathers and swimmers in the shallow Gulf waters.

Dangerous Waves

Waves, in addition to causing rip currents, can be dangerous, especially when breaking onshore. Shorebreaks, which occur when large waves break directly on the beach face, are generally the most dangerous. Only very infrequently do rogue or "sleeper waves" occur, and modern technology is quite effective in warning us of impending tsunamis, which are mistakenly called tidal waves.

Steep beaches, where plunging waves break in shallow water, are subject to shorebreak (figure 11). Ocean swells that hit these beaches often pass from fairly deep water abruptly into very shallow water, causing the waves to rise up quickly in height and to break with considerable downward force. A good example of shorebreaks is at Sandy Beach on the island of Oahu, Hawaii. Waves five to eight feet high are known to break in knee-deep water during heavy surf conditions. Even so, Sandy Beach is a popular recreational beach in Hawaii, because some people enjoy being tossed and rolled around by the big waves. For daring swimmers, shorebreaks can provide exciting waves for bodysurfing, but this activity is dangerous, and even so-called experts can be seriously injured. Broken necks, which can cause paralysis or death, are of major concern.

Fortunately, shorebreaks are not common on most U.S. beaches and are virtually unknown along the Gulf of Mexico's gently sloping beaches. Waves there normally break on a sandbar, so that a broken wave plunges into the deeper water just landward of the underwater bar before running up the beach face as swash. Anyone caught in such a breaker would be

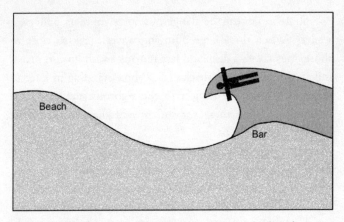

Figure 11. Shorebreak is a dangerous wave that breaks directly on the beach with great force.

thrown into a cushion of water rather than being picked up and driven head first into the sand by a shorebreak.

Waves often come in groups, called surf beat, that crescendo up to the really big breakers. Surfers often wait out the smaller waves to ride the biggest waves of the set, which are usually the seventh to tenth waves along some Southern California beaches. These larger plunging breakers are a surfer's delight but can imperil swimmers or bathers who are caught unaware. The waves can seem manageable until the set rolls in, causing dangerous shorebreaks. The trick is to take a big breath of air and to dive *under* the incoming waves. Inexperienced swimmers and bathers will attempt to jump through or over these breakers, which invites disaster. You can be pounded head first into the hard-packed sand beach, which can have the same result as jumping head first into a shallow pool.

Shorebreaks can be dangerous for swimmers and even waders. When these large plunging waves break so close to shore, they create a deep and fast-moving swash uprush that is followed by a powerful backwash. People standing in knee to waist-deep water are subject to the vigorous suction of the backwash, pulling them seaward toward the next incoming breaker.

Backwash is simply the return of the water from the broken wave (swash) that moves up the beach face. I really enjoy walking along the shore in the swash zone as the water advances and recedes. When the waves are large and powerful, the uprush of the swash can produce a whitewater experience for board and body surfers. While some of the water seeps into the beach, a big swash uprush means a strong return flow of the backwash on a steeply inclined beach, and it can be swift and deep enough to prohibit escape. The returning flow can knock your feet out from under you and drag you into deeper water. Most people mistakenly think that the backwash is the mythical undertow because of the strong sucking action, but the backwash stops at the next breaking wave and does not go beyond the surf zone, as a rip current does. Still, the results can be life-threatening during times of high surf as the waves crash down on you, causing injury by shorebreak or drowning from swallowing too much salt water.

There is an old Hawaiian saying: "Never turn your back to the sea." The Pacific Ocean beaches, especially in Hawaii and on the U.S. northwest coast, are subject to "sleeper" or rogue waves.

These single large waves appear to come out of nowhere, and unsuspecting victims have been swept off of cliffs or crashed into rocks to certain death. These waves are different from surf beat, where the surf gradually crescendos as the larger waves in a set roll ashore. Rogue waves are generally caused by two different wave trains coming from different directions and their crests peaking at the same time. The resulting breaker, which is both preceded and followed by regular-size waves, can be surprisingly large—twice as high as normal. Such waves are rare on sandy swimming beaches, but many lives have been lost on rocky coasts, where deep water is close to shore.

The largest waves that can strike a beach are tsunamis. These walls of water, which can reach 50 feet in height, are instead caused by violent movements of the sea floor, usually triggered by underwater earthquakes. The geologically active Pacific coast, and the Caribbean (to a lesser extent), are subject to tsunami waves. These waves cannot be spotted at sea in deep water, contrary to the conditions portrayed in the Hollywood movie *The Poseidon Adventure*. Instead, these long-period waves bunch up as they approach shallow water to reach tremendous heights upon breaking. There are tsunami warning systems throughout the Hawaiian Islands, which have experienced a number of these giants throughout history.

Conclusion

Beaches are the number one recreational destination for Americans, and people are flocking to the shore in ever-increasing numbers. Nothing restores the body and soul like a stay at the beach. We are naturally drawn to the rhythmic pounding of the waves, as if returning to our primordial beginnings.

The goal of this book is to ensure that your beach experience fulfills your expectations and is not marred by injury or, worse, tragedy.

As you've read, sharks are rarely a problem on swimming beaches, despite the general uneasiness that the movie *Jaws,* along with the media hype over recent attacks, has instilled in some people. Far more people are killed by rip currents, which are the biggest threat to bathers and swimmers. The ratio of shark bites to rip rescues is approximately 1 to 10,000—this staggering statistic puts things into proper perspective. For instance, in the year 2000 there were 70,771 rescues from drowning and only 54 unprovoked shark attacks (and no deaths) in U.S. coastal waters.[1]

My travels have taken me to all the major recreational beaches in the United States and many worldwide, and I always look forward to my next trip. I hope that your future journeys to the beach will be safe and enjoyable.

Shark Facts

- Worldwide 70 to 100 shark attacks occur each year; only 15 percent are fatal.
- In California there is only one shark attack for every million surfers.
- Almost 90 percent of shark attacks in American waters occur at the surface or less than five feet from the surface. This shark fact can be misleading—most attacks are not right at the beach. Surfers are the number one attack victims.
- Sharks—the "steely-eyed fish from hell"—have been swimming in the world's oceans for more than 450 million years, which is 200 million years before the dinosaurs roamed the earth.
- Sharks have keen senses; they can hear a wounded animal struggling in the water more than a mile away and can smell a single drop of blood in the water column several thousand feet away.
- Sharks often lose teeth when they eat and can go through more than 20,000 teeth in a lifetime.
- Great whites are the largest carnivorous sharks; the largest recorded great white shark was 21 feet long and weighed greater than 4,000 pounds.

- Pygmy sharks have a maximum length of 10 inches and weight of 1 ounce.
- Whale sharks reach almost 40 feet long with a weight of 14 tons, but are harmless to people.

Rip Current Statistics

- Eighty percent of ocean rescues (more than 70,000 per year) involve saving someone caught in a rip current.
- A strong rip current moves at 3 feet per second, which is as fast as an Olympic swimmer.
- Rip currents have been measured with speeds as high as 5 feet per second at big surf beaches in Australia.
- Rips often occur at groins, jetties, and piers; stay at least 100 feet away from these structures to avoid these deadly currents and other hazards.
- About 70 to 100 people drown annually in rip currents in the United States; 20 to 30 of them occur in Florida because of its large number of good swimming beaches and generally pleasant weather year-round.[2]
- Breaking waves that approach 5 feet can generate powerful rip currents. The energy of a wave is proportional to its height; thus a 3-foot wave is 9 times more powerful than a one-footer.

Notes

Introduction

1. *Baltimore Sun,* September 1, 1995, "Storms bring deadly energy to surf."

one**Sharks**

1. www.njhm.com, November 6, 2001.
2. *U.S. News and World Report,* April 17, 1998.
3. *Santa Cruz Sentinel,* September 30, 2000, "Maverick's surfer escapes shark attack with minor cut."
4. *San Francisco Chronicle,* August 28, 1998, "Shark experts don't see need to close beach for long; attack was ninth along Marin coast."
5. *National Geographic,* April 2000, "Inside the Great White."
6. *Maui News,* August 14, 2000, "Sharks are gone; park open again."
7. *Time,* July 30, 2001, "Summer of the Shark."
8. Ibid.
9. Ibid.
10. *New York Times,* September 2, 2001, "Boy dies after shark attack in Virginia."

11. *New York Times,* September 4, 2001, "Man killed in North Carolina shark attack."
12. *St. Petersburg Times,* September 1, 2000, "Splash triggered feeding shark to attack."
13. George Burgess, director, International Shark Attack File.
14. Ibid.
15. *Miami Herald,* August 26, 2001, "Surfer 9th shark-bite victim in week."
16. Stevens, *Sharks.*
17. Ibid.
18. Ibid.
19. *Time,* July 30, 2001, "Summer of the Shark."
20. Stevens, *Sharks.*
21. Ibid.
22. *Miami Herald,* August 16, 2001, "Debate surfaces over attack on swimmer, shark-adventure dives."

two**Rip Currents**

1. *New York Times,* July 24, 2001, "3 girls, trapped in surf at Rockaways, drown."
2. Jim Lushine, National Weather Service, Miami.

Conclusion

1. Florida Sea Grant, "Sharks in Perspective: From Fear to Fascination," Chris Brewster and George Burgess, June 12–14, 2002.
2. Jim Lushine, National Weather Service, Miami.

For Further Reading

Baldridge, H. D. *Shark Attack*. New York: Berkley, 1974.

Bascom, W. *Waves and Beaches*. New York: Doubleday, 1980.

Benchley, Peter. *Shark Trouble*. New York: Random House, 2002.

Brewster, B. C. *The United States Lifesaving Association Manual of Open Water Lifesaving*. Englewood Cliffs, N.J.: Prentice Hall, 1985. 316 pp.

Burgess, G. H., and C. S. Macfiel. "Encounters with Sharks in North and Central America," in *Sharks,* J. D. Stevens, ed. New York: Prentice Hall, 1999.

Clark, E. *The Lady and the Sharks*. New York: Harper & Row, 1969.

Clark, J. R. K. *Hawaii's Best Beaches*. Honolulu: University of Hawaii Press, 1999. 148 pp.

Cousteau, J. Y. *The Shark: Splendid Savage of the Sea*. London: Cassell, 1989.

Ellis, R., and J. E. McCosker. *Great White Shark*. New York: Harper-Collins, 1991.

Griffiths, T., *Better Beaches: Management and Operation of Safe and Enjoyable Swimming Beaches*. Hoffman Estates: National Recreation and Park Association, 1999. 148 pp.

Leatherman, S. P. *America's Best Beaches*. Gainesville: University Press of Florida, 1998. 112 pp.

Lushine, J. B. "Rip Currents: Human Impact and Forecastability," *Coastal Zone* 91 (1991): 3558–3569.

Short, A. D., and C. L. Hogan. "Rip Currents and Beach Hazards: Their Impact on Public Safety and Implications for Coastal Management," *Journal of Coastal Research* 12 (1994): 197–209.

Sonu, C. J. "Field Observations of Nearshore Circulation and Meandering Currents." *Journal of Geophysical Research* 77 (1972): 3232–3247.

Stevens, J. D. *Sharks.* New York: Checkmark Books, 1999. 240 pp.

Wilson, Edward O. *The Diversity of Life.* New York: Norton, 1992.

Relevant Websites

Healthy Beaches Campaign: *www.DrBeach.org*

International Shark Attack File: *www.flmnh.ufl.edu/natsci/ichthyology/shark.htm*

Mote Marine Laboratory: *www.mote.org*

National Oceanic & Atmospheric Administration: *www.nmfs.noaa.gov*

Natural Resources Defense Council: *www.nrdc.org*

Ocean Conservancy: *www.oceanconservancy.org*

Shark Research Institute: *www.sharks.org*

Surfrider Foundation: *www.surfrider.org*

United States Environmental Protection Agency: *www.epa.gov/OST/beaches/*

United States Lifesaving Association: *www.usla.org*

Appendixes

A. Shark Attacks Versus Other Injuries

Shark Attacks Versus Home Improvement Injuries in the U.S. in 1996

Cause of injury	Number of injuries
Nails, screws, tacks, and bolts	198,849
Ladders	138,894
Toilets	43,687
Buckets	10,907
Sharks	18

Source: Florida Marine Research Institute

Shark Attacks Versus Lightning Strikes

	Period	Number of lightning injuries	Number of lightning fatalities	Number of shark attacks	Number of fatal shark attacks
Florida	1959–1994	1,178	345	180	4
U.S.	1959–1994	9,818	3,239	336	12

Source: National Weather Service

Shark Attacks Versus Other Animal Fatalities Worldwide

Category	Annual average
Shark-related fatalities	10
Elephant-related fatalities	250
Bee-sting fatalities	1,250
Alligator-related fatalities	2,500

B. Warning Signs on the Beach

HIGH SURF

STRONG CURRENT

NO SWIMMING

WAVES ON LEDGE

SUDDEN DROP OFF

NO BOARDSURFING

DANGER SHOREBREAK

SLIPPERY ROCKS

NO DIVING

MAN-OF-WAR

SHARP CORAL

NO BOARDSAILING

Source: U.S. Lifesaving Association

C. Common Marine Organisms That Can Injure You

Organism	Common habitat	Injury prevention	Injury mechanism	Symptoms	First aid
Coral	Tropical waters	Do not walk on reefs or handle coral.	Sharp edges	Lacerations and skin abrasions.	Clean cut thoroughly, removing any coral that has broken off in the wound. Bandage and keep clean.
Jellyfish, especially Portuguese man-of-war	Oceanic surface waters	Avoid swimming in areas with jellies present.	Tentacles	Stinging, swelling, welts, and redness of the affected area. Severe reactions include difficulty breathing and cardiac arrest.	*Jellyfish:* Apply vinegar to the affected areas. Apply ice for pain. If reaction is severe, seek immediate medical attention. *Men-of-War:* Remove any visible tentacles. Rinse in fresh or salt water. Apply ice for pain. Immediate medical attention may be required for severe reactions.

Moray eel	Coral reefs or other rocky areas	Avoid probing crevices with your hands. Give a wide berth to swimming eels.	Powerful jaws and sharp teeth	Lacerations, including muscle, nerve, and bone damage.	Control bleeding and seek medical attention.
Sea urchin	Shallow reef waters	Use protective footwear and do not handle.	Spines	Skin discoloration and throbbing pain in affected area.	Apply pressure to control bleeding; apply vinegar to the affected areas.
Shark	All oceanic waters	Avoid swimming at dawn, dusk, or in murky water. Exit water if a shark is spotted.	Powerful jaws and rows of sharp teeth	Severe lacerations, including damage to muscles, nerves, bones and/or loss of limbs.	Control bleeding and seek immediate medical attention.
Sting ray	Sandy shallows to deep ocean waters	Do not approach swimming rays; shuffle feet when walking in shallow waters.	Poisonous spine on tail	Lacerations, severe pain, nausea, and development of tetanus.	Apply pressure to control bleeding; and seek medical attention. Heat (e.g., hot water) takes pain away.

Note: These are the well-known first-aid treatments. For further treatment, consult your physician.

D. America's Top Beaches

Each year since 1991, I have listed the top American beaches as determined by fifty specific criteria (see appendix E). Each year's winner is not eligible for ranking in subsequent years. Following is a list of all the number-one ranked beaches, along with the runners-up in 2002.

1991   Kapalua Bay Beach   Maui, Hawaii
Kapalua is the perfect beach for swimming and snorkeling—its calm, clear waters are full of colorful tropical fish. This crescent-shaped pocket beach of white coral sand is lined with palm trees, making for an idyllic setting.

1992   Bahia Honda State Recreation Area   Florida
The Florida Keys are known best for their diving and snorkeling, but Bahia Honda Key boasts two wonderful beaches: Sandspur Beach and Caloosa Beach. This is really a piece of the Caribbean, and the whole island is designated a state park.

1993   Hapuna Beach State Recreation Area
Big Island, Hawaii
Hapuna is an oasis of white sand and palm trees surrounded by black volcanic lava; it is difficult to believe that such a beautiful beach can exist in such a barren environment. The Big Island is the only place in America where you can snow ski and water ski on the same day.

1994   Grayton Beach State Recreation Area   Florida

The Florida panhandle has the finest, whitest sand in the world. The crystalline quartz sand here looks like sugar, and the emerald-green waters are perfectly clean and clear. Beach development has been restrained, so that the natural landscape, with its big sand dunes, still dominates.

1995   St. Andrews State Recreation Area   Florida

St. Andrews has all the amenities of nearby Panama City and yet the beauty of a natural landscape. This is one of the best places in the country for bird watching and beachcombing for seashells. The water and sand are fantastic along this famed panhandle coast.

1996   Lanikai Beach   Oahu, Hawaii

Lanikai Beach is picture perfect, with two islands framing the clear emerald-green water. All the beaches in Hawaii are public, but you must search for a parking place to explore this small white-coral-sand beach, which is a real treasure.

1997   Hulopoe Beach   Lanai, Hawaii

This long, crescent-shaped beach of white coral sand is framed by the nearby mountains in the tropical paradise of Hawaii. Lanai is the highest and one of the most remote islands in the chain; it is where the locals go on vacation.

1998   Kailua Beach Park   Oahu, Hawaii

Kailua is characterized by calm, clear water and very fine sand. It is an excellent place for water sports, such as sea kayaking, or for just taking a jog or walk. This beach park offers all the amenities in the small seaside town of Kailua.

1999   Wailea Beach   Maui, Hawaii

Wailea is located on the sunny, dry side of Maui, making every day a beach day. This coast has five small pocket beaches, each separated by lava flows and backed by beautiful gardens and tropical trees (and some fantastic seaside hotels).

2000   Mauna Kea Beach   Big Island, Hawaii

Mauna Kea (also known as Kaunaoa Beach), with sparkling clear water, is one of the best places to spot the rare Hawaiian green sea turtles and many other sea creatures.

2001   Poipu Beach Park   Kauai, Hawaii

Poipu Beach Park is one of the best swimming areas in Hawaii. Its golden coral sands jut out into the water and are anchored by an island, forming a tombolo-shaped beach. This makes for protected, calm water, which is great for snorkeling with the tropical fish (my favorite activity). Where the beach is unpro-tected from the open Pacific, visitors can ride the big waves on surfboards.

2002   St. Joseph Peninsula State Park   Florida

St. Joe projects far out into the clear and clean aquamarine waters of the Gulf of Mexico. The surf is low, making for excellent swimming in a peaceful environment dominated by tall sand dunes and varied wildlife, including hawks and foxes. Rounding out the top nine in 2002 are:

2. Hanalei Beach, Hawaii
3. Kaanapali, Hawaii
4. Fort DeSoto Park, Florida
5. Caladesi Island State Park, Florida
6. Ocracoke Island, North Carolina
7. Hamoa Beach, Hawaii
8. East Hampton Beach, New York
9. Cape Florida State Recreational Area, Florida
10. Hanauma Bay, Hawaii

E. How the Top Beaches Are Chosen

Stephen P. Leatherman's fifty criteria for ranking American beaches

PHYSICAL FACTORS (relate to the vacation/ holiday season)	CATEGORIES				
	1	2	3	4	5
1. Beach width at low tide	<10 m	10–30 m	30–60 m	60–100 m	>100 m
2. Beach material	cobbles	sand/cobbles	coarse sand	—	fine sand
3. Beach condition or variation	erosional	—	stable	—	depositional
4. Sand softness	hard	—	—	—	soft
5. Water temperature	cold/hot	—	—	—	warm (70°–85° F)
6. Air temperature (midday)	< 60° F >95° F	—	—	—	80°–90° F
7. Number of sunny days	few	—	—	—	many
8. Amount of rain	large	—	—	—	little
9. Wind speeds	high	—	—	—	low
10. Size of breaking waves	high/dangerous	—	—	—	low/safe
11. Number of waves/ width of breaker zone	none	1–2	3–4	5	6+
12. Beach slope (underwater)	steeply sloping bottom	—	—	—	gently sloping bottom
13. Longshore current	strong	—	—	—	weak
14. Rip currents	often	—	—	—	never
15. Color of sand	gray	black	brown	light tan	white/pink
16. Tidal range	large (>4 m)	3–4 m	2–3 m	1–2 m	small (<1 m)
17. Beach shape	straight	—	—	—	pocket
18. Bathing area bottom conditions	rocky, cobbles, mud	—	—	—	fine sand

PHYSICAL FACTORS (relate to the vacation/ holiday season)	CATEGORIES				
	1	2	3	4	5
19. Turbidity	turbid	—	—	—	clear
20. Water color	gray	—	—	—	aqua blue
21. Floating/suspended human material (sewage, scum)	plentiful	—	—	—	none
22. Algae in water (amount)	infested	—	—	—	absent
23. Red tide	common	—	—	—	none
24. Smell (e.g., seaweed, rotting fish)	bad odors	—	—	—	fresh salty air
25. Wildlife (e.g., shore birds)	none	—	—	—	plentiful
26. Pests (biting flies, ticks, mosquitoes)	common	—	—	—	no problem
27. Presence of sewage/ runoff outfall lines on or across beach	several	—	—	—	none
28. Seaweed/jellyfish on the beach	many	—	—	—	none
29. Trash and litter (paper, plastic, nets, ropes, planks)	common	—	—	—	rare
30. Oil and tar balls	common	—	—	—	none
31. Glass and rubble	common	—	—	—	rare
32. Views and vistas —Local scene	obstructed	—	—	—	unobstructed
33. Views and vistas —Far vista	confined	—	—	—	unconfined
34. Buildings/urbanism	overdeveloped	—	—	—	pristine/wild
35. Access	limited	—	—	—	good

(*continues*)

Stephen P. Leatherman's fifty criteria for ranking
American beaches (*Continued*)

PHYSICAL FACTORS (relate to the vacation/ holiday season)	CATEGORIES				
	1	2	3	4	5
36. Misfits (nuclear power station, offshore dumping)	present	—	—	—	none
37. Vegetation nearby (trees, sand dunes)	none	—	—	—	many
38. Well-kept grounds/ promenades or natural environment	no	—	—	—	yes
39. Amenities (showers, chairs, bars, etc.)	none	—	—	—	some
40. Lifeguards	none	—	—	—	present
41. Safety record (deaths)	some	—	—	—	none
42. Domestic animals (e.g., dogs)	many	—	—	—	none
43. Noise (cars, nearby highways, trains)	much	—	—	—	little
44. Noise (e.g., crowds, radios)	much	—	—	—	little
45. Presence of seawalls, riprap, concrete/rubble	large amount	—	—	—	none
46. Intensity of beach use	overcrowded	—	—	—	ample open space
47. Off-road vehicles	common	—	—	—	none
48. Floatables in water (garbage, toilet paper)	common	—	—	—	none
49. Public safety (e.g., pickpockets, crime)	frequent problems	—	—	—	infrequent problems
50. Competition for free use of beach (e.g., fishermen, boaters, water-skiers)	much	—	—	—	little